D1779878

Marine Geology of Korean Seas

Marine Geology of Korean Seas

Sung Kwun Chough
Seoul National University

International Human Resources Development Corporation • Boston

Cover design by Diane Sawyer

Interior design by Susan Swanson

©1983 by International Human Resources Development Corporation. All rights reserved. No part of this book may be used or reproduced in any manner whatsoever without written permission of the publisher except in the case of brief quotations embodied in critical articles and reviews. For information address: IHRDC, Publishers, 137 Newbury Street, Boston, MA 02116.

Printed in the United States of America

Library of Congress Cataloging in Publication Data

Chough, Sung Kwun, 1946—
  Marine geology of Korean seas.

  Bibliography: p.
  Includes index.
  1. Geology—Korea. 2. Geology—Yellow Sea.
3. Geology—Japan, Sea of. I. Title.
QE305.C48 1983   551.46′554   83-8512
ISBN 0-934634-61-0

Geological Sciences Series

Series Editor
George deVries Klein
Department of Geology
University of Illinois at Urbana-Champaign

Consulting Editors
Michael A. Arthur
Department of Geology
University of South Carolina

Thomas W.C. Hilde
Department of Geophysics
Texas A&M University

W. Stuart McKerrow
Department of Geology and Mineralogy
University of Oxford

J. Casey Moore
Earth Sciences
University of California-Santa Cruz

To my Parents

Contents

Preface xi

Acknowledgments xiii

1   Introduction   1
        Previous Work   1
        Korean Peninsula   7

2   *Epicontinental Sea (Yellow Sea)*   15
        Physiography   15
        Geologic Structure   15
        Shallow Structure   27
        Surface Sediments   36
        Dispersal Pattern   44

3   *Marginal Sea (East Sea)*   55
        Geologic Setting   55
        Velocity Structure   58
        Magnetic and Gravity Anomalies   61
        Heat Flow   63
        Acoustic Stratigraphy   63
        Surface Sediments   68
        Late Quaternary Stratigraphy   71
        Paleoceanography   74

4   *Eastern Shelf*   77
        Geologic Setting   77
        Acoustic Stratigraphy   77
        Surface Sediments   84

5   *Ulleung Basin*   89
        Physiography   89
        Acoustic Stratigraphy   89
        Mass-Flow Deposits   92
        Turbidite Facies   97
        Hemipelagic Facies   100
        Stratigraphy   102
        Turbidite Provenance   103

6   *Coastal Embayments (South Coast)*   111
    General Statement   111
    Gamagyang Bay   111
    Deugryang Bay   122
    Late Quaternary History   128

7   *Geological History*   131
    General Statement   131
    Precambrian   131
    Paleozoic   132
    Mesozoic   133
    Cenozoic   134
    Late Quaternary   136

References   139

Addendum   151
Block 4   151
Subzone 5   151
Subzone 7   151
Block 6   152
Additional References   152

Index   153

Preface

The shallow and deep seas bounding the Korean Peninsula have not only aroused the curiosity of the Korean people who believe that the seas would bring prosperity, but have stimulated scientific investigations as well.

In the last two decades, two decades later than the pioneering postwar endeavours of U.S. and European countries, a group of Korean geologists have begun to explore the sea floor and its subsurface with elementary equipment. This group was led by the sea-going research teams of the Korea Institute of Energy and Resources (KIER; formerly Geological Survey of Korea, Geological and Mineral Institute of Korea, and Korea Research Institute of Geosciences and Mineral Resources). To map the sea floor and identify sedimentary strata beneath it, the teams covered a relatively small area each year focusing on the shallow structure (usually less than 80 m) and characteristics of bottom sediments within about 80 km from the shoreline.

Marine geologists in academic institutions and the government-supported research institute, Korea Ocean Research and Development Institute (KORDI, established in 1973), also took a role in carrying out studies in cooperation with the KIER teams. A number of joint ventures with advanced, foreign and international organizations have also been made during this time for gathering basic geological information in the seas aimed at exploring mineral and hydrocarbon resources. In the meantime, the East Sea (Sea of Japan) has been the target of numerous investigations by American, Japanese, Russian, and recently Korean geologists.

Many interesting papers have been written by Korean geologists (some with foreign geologists), mostly in domestic journals or as unpublished manuscripts. These geologists, however, are seldom recognized by others who are interested in the geology of the seas around the Korean Peninsula. This appears to be an appropriate time for a summary of the Korean Seas because the available information is adequate now for some generalizations.

This monograph was conceived in the autumn of 1981. It was written not only as a reference book but also as a guide to graduate students in a course on marine geology in the Department of Oceanography at the Seoul National University (SNU).

For this purpose, I have focused on the geologic structure, sedimentary facies and geological history of the seas, and on the geological processes that operate in them.

This book is not a treatise on every aspect of marine geology of Korean Seas such as nearshore and estuarine processes. Rather it is confined to some facts or problems recognized by data on the offshore. Physical parameters such as currents, temperature, and salinity in general, are not treated unless needed in discussing some recent processes of sediment transport and deposition. Because of the lack of our knowledge on the northern part of the peninsula (north of 38th parallel; DMZ), this synthesis stresses only the sea floor off the Republic of Korea.*

---

*Geographic names that appear in the text mostly follow the present-use grammar, a system proposed by the Ministry of Education, represented by the Korean standard of English pronunciation. Former notations and names that have been misspelled or incorrect are, in most cases, shown in parentheses.

Acknowledgments

I would like to thank the following publications and copyright holders for their cooperation: Report and Bulletin of the Korea Institute of Energy and Resources (Korea Institute of Geosciences and Mineral Resources, Geological and Mineral Institute of Korea, Geological Survey of Korea) (figs. 1.3, 1.4b,c, 2.2, 2.10, 2.11, 2.15, 2.16, 2.17, 4.2, 4.3, 4.7, 6.9, 6.10); Journal of Sedimentary Petrology (figs. 2.1, 2.25, 2.26, 5.5, 5.7, 5.8, 5.9); Marine Geology (figs. 3.7, 6.1, 6.3, 6.4, 6.5, 6.6, 6.7); American Association of Petroleum Geologists Bulletin (figs. 2.4, 2.8, 2.9); Bulletin of Committee for Co-Ordination of Joint Prospecting for Mineral Resources in Asian Offshore Areas (figs. 2.5, 2.6, 2.23, 2.24, 4.1); Bulletin of Korea Ocean Research and Development Institute (figs. 2.3, 2.21, 6.2); Chinhae Machine Depot (fig. 4.8); Bulletin of the Geological Society of America (figs. 1.1, 2.19, 3.3, 3.4); Oceanus (fig. 3.2); Bulletin of Geological Survey of Japan (figs. 3.8, 4.6); U.S. Naval Oceanographic Office (fig. 4.4); American Geophysical Union (figs. 3.6, 7.1); and Journal of Marine Geophysical Research (fig. 3.5).

I am indebted to many colleagues in Seoul and abroad for invaluable dialogue and support, especially C.S. Kim and the members of the Marine Geology and Geophysics Divisions, KIER (K.J. Cho, S.W. Kim, J.H. Chang, H.C. Han, C.M. Kim, W.S. Kim, C.W. Lee, W.Y. Lee, Y.O. Lee, G.H. Min, K.P. Park, K.S. Park, and W.C. Shin) and the colleagues at the Marine Geology Laboratory, KORDI (S.K. Chang, S.J. Han, K.S. Jeong, and B.C. Suk) who generously permitted me the use of unpublished data and help in acquiring data and preparing illustrations. Discussions with them, as well as with M.S. Lee, B.K. Park, K.H. Chang, B.K. Kim, C.H. Cheong, S.M. Lee, Y.H. Kwak, K.H. Paik, H.I. Choi, S.J. Kim, J.K. Oh, J.H. Oh, J.K. Park, Y.S. Kong, and M.Y. Song were useful in clarifying points in their field of interest. G.deV. Klein and D.S. Gorsline read the entire manuscript and made helpful suggestions. K.O. Emery also offered valuable comments for the improvement of the manuscript. E. Honza, E. Inoue, and K. Tamaki of the Geological Survey of Japan provided valuable information through cooperative research. I thank G.deV. Klein and M. Hays for their efforts for successful publication of this work. I am indebted to both the Korea Science and Engineering Foundation and the Research Institute of Basic Sciences, Seoul National University for their continuous support through research grants.

I am grateful to my colleagues (Y.A. Park, J.Y. Chung, J.H. Shim, K. Kim, C.H. Koh, and C.B. Lee) and graduate students in the Department of Oceanography who have carried out marine geological and oceanographic studies on the Korean Seas. The editorial aid of K.S. Bahk, M.Y. Choi, H.J. Kang, and C.P. Yoder was invaluable. Most of the figures were drawn by H.J. Kim and his team at the Instructional Media Center (SNU). J.S. Cho typed the manuscript. All of their help is gratefully acknowledged.

I am grateful to my wife for her encouragement over the years of my research and writing.

Marine Geology of Korean Seas

1 Introduction

Previous Work

The Korean Seas (fig. 1.1) are geologically unique. The Yellow Sea (or West Sea) is a shallow (less than about 100 m), postglacially submerged epicontinental sea bound on the east by a long stretch of ria-type coast. The East Sea, the western part of the Sea of Japan, is characterized by a narrow shelf with a straight coastline. The Yellow Sea floor is rather flat and progressively deepens toward the southeast to form the Okinawa Trough in the northern East China Sea. The East Sea deepens abruptly seaward, forming a number of deep basins between ridges and surrounding margins that are related probably to the rifting of a back-arc basin associated with the subduction of the Pacific Plate. The South Sea, bounding the southern coast of the peninsula, is also shallow and flat, similar to the Yellow Sea, but characterized mostly by rocky embayments.

Regional studies on the geological structure of the Yellow Sea were made in a joint survey (Emery et al. 1969; C.S. Kim et al. 1969) supported by the Committee for Co-Ordination of Joint Prospecting for Mineral Resources in Asian Offshore Areas (CCOP) (fig. 1.2). An airborne magnetic survey was also conducted in the Yellow and South seas and the southern part of the East Sea (Bosum et al. 1971). These studies showed in the Yellow Sea the existence of a number of Tertiary continental basins with deformed strata, whose sedimentation was controlled by a series of folded basement ridges at the mouth of the Yellow Sea (Emery et al. 1969).

A regional unconformity of Plio-Pleistocene age was also found overlying the deformed Tertiary strata. The unconformity is progressively shallower toward the Korean Peninsula (Frazier et al. 1976), and bears an important implication for the paleoceanography and paleoenvironment in the sea. It has yet to be traced near the peninsula. Attempts have been made by the KIER mapping projects since the early seventies to obtain data on the geological structure of the shallow portions of the Yellow Sea (e.g., Koo et al. 1971; C.S. Kim et al. 1972; C.M. Kim and Lee, 1974a, 1974b; C.S. Kim 1976; S.W. Kim et al. 1980a, C.M. Kim et al. 1981).

Closely spaced deep seismic reflection profiles and a number of exploratory boreholes were obtained in the Yellow and East seas

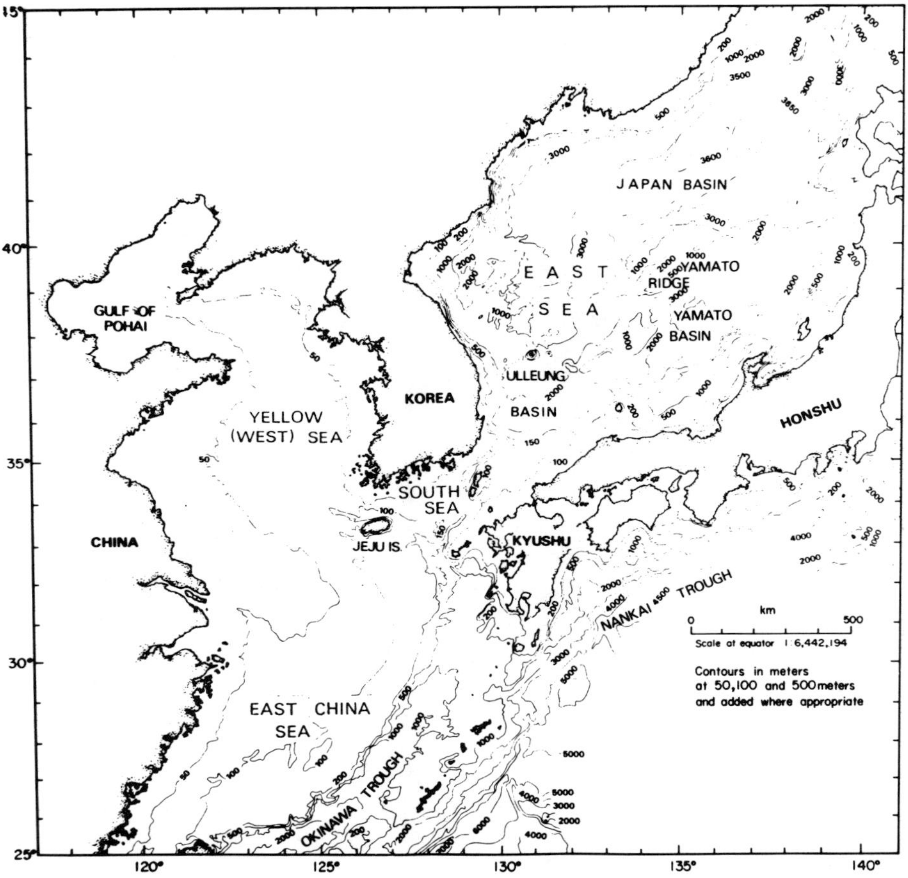

Figure 1.1. Bathymetry of the Korean Seas: the Yellow Sea, northern part of the East China Sea, and the East Sea (Sea of Japan). Contours in meters. Modified after Mammerickx et al. (1976) by permission of the Geological Society of America.

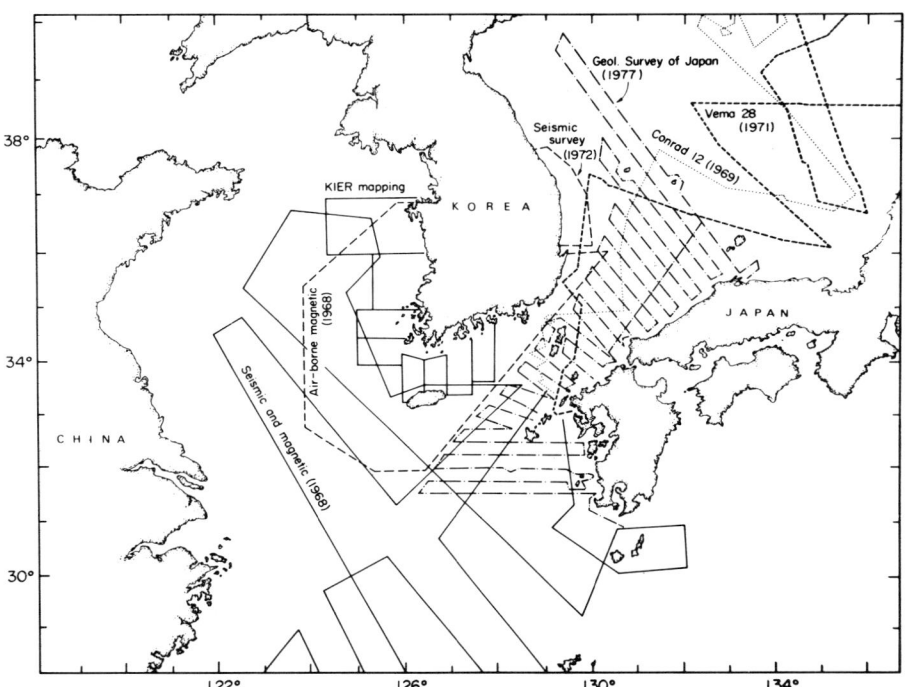

Figure 1.2. Tracklines and blocks of major geological and geophysical surveys in the seas around the Korean Peninsula.

(fig. 1.3) by a number of foreign oil companies under contract, generally confirming the data revealed earlier. Drilling activities have been greatly enhanced recently in the South Sea and northern East China Sea aiming at the deformed Tertiary strata whose economic potential for hydrocarbon was strongly predicted by the earlier studies. As early as 1961, Niino and Emery described the surface sediment distribution in the entire Yellow Sea and East China Sea by compiling and analyzing the data obtained prior to 1940.

In the continental margin of the East Sea, the Geological and Mineral Institute of Korea (presently KIER) conducted a cooperative seismic survey with the Federal Institute of Geoscience and Mineral Resources of West Germany (Schlüter and Chun 1974) to reveal possible seaward extensions of coal seams in the northern part (below 38th parallel) and hydrocarbon potential off Pohang (fig. 1.2). A high resolution seismic study was also made in the latter area by Huntec (1968) under the support of the United Nations Economic Commission for Asia and the Far East (ECAFE) and the United Nations Development Programme (UNDP). Structurally, the margin consists of north-south trending, pre-Tertiary horst-type embankments and basins (K.P. Park et al. 1981). The data revealed thus far indicate that the East Sea margin was formed in the early stage by block faulting related probably to the rifting of the Ulleung back-arc basin.

Although the origin of East Sea (Sea of Japan) is still speculative, the geologic structure of the sea has been discussed by many American, Japanese, and Russian geologists (e.g., Beresnev and Kovylin 1970; Bersenev 1971; Ludwig et al. 1975; Kobayashi and Isezaki 1976; Gnibidenko 1979; Uyeda 1979). There are essentially two hypotheses on the origin of the sea. One hypothesis concerns back-arc spreading by lateral, southeastward migration of the Japanese island arc. This hypothesis is favored by American and most Japanese and Korean geologists. The other hypothesis assumes a vertical movement of a former landmass to its present depth by oceanization, a view proposed by most Russian and some Japanese workers. It is, however, beyond the scope of this book to discuss the origin of the entire sea. Rather the book will deal with the facts related to the tectonics and sedimentary facies in

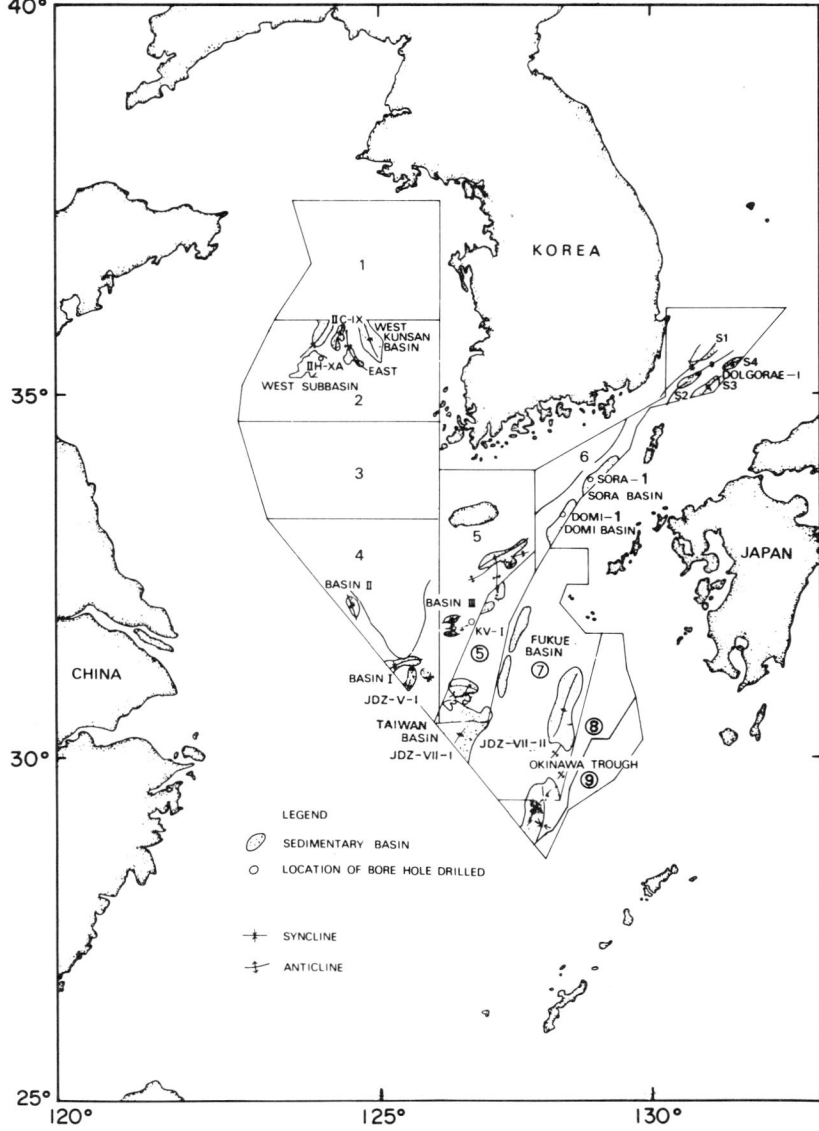

Figure 1.3. Concession blocks (1 to 6) for hydrocarbon exploration in the Korean Seas. Circled numbers for the subzones of the Korea-Japan Joint Development Zone (JDZ). Target sedimentary basins are also shown. Figure courtesy of the Korea Institute of Energy and Resources.

the western portion of the sea, such as the Ulleung (Tsushima) Basin, Korea Plateau, and the associated eastern margin of the Korean Peninsula.

The Ulleung Basin, whose eastern two-thirds were surveyed extensively by Ludwig et al. (1975) and Honza (1979a, 1979b), was found to consist of oceanic basement overlain successively by transparent, hemipelagic sediments, and turbidites. The turbidite facies, composed mainly of fine-grained parallel laminated mud (Chough 1982), has been predominant probably since late Pliocene. The study of turbidites in the basin has been enhanced recently, also revealing that the extensive turbidite facies in the basin is transitional to the various types of submarine mass flow deposits such as slump, slide, and debris flow facies on the surrounding slope. The basin is also bounded on the north by the Korea Plateau, which is a probable extension of the Jurassic and Cretaceous granites and poses an important criterion for geologic information on land and under the sea. Thick turbidite sediment and high heat flow in the basin are suggestive of hydrocarbon potential and these deposits may become targets for future exploration.

The southern coast of the peninsula (fig. 1.1) is characterized mostly by postglacially submerged rocky embayments. Sub-bottom reflection surveys in the embayments (J.H. Chang et al. 1980; H.J. Kang and Chough 1982) indicate that the basement (mostly Cretaceous volcanics) is overlain by a thick (up to 30 m) Quaternary sediment sequence. Strong mid-reflectors in the sedimentary sequence indicate that the southern coastal sea and embayments have alternately been submerged and emergent during the Quaternary period. In the postglacial period, fine-grained sediments transported from offshore were trapped in the embayments, a unique environment of geologic interest. Sediments were often found to contain greater volumes of organic carbon. Geological processes related to the postglacial sea-level changes proved to be important along the coast of the peninsula, both for the Quaternary history (Y.A. Park 1969) and for the problems related to the human activity in land use. Efforts to find economic minerals in the offshore and nearshore areas prompted recently the KIER and KORDI marine geology teams to obtain and evaluate geochemical and mineralogical data (Y.O. Lee 1979; Koo et al. 1980b).

Many geological aspects that emerged for the studies outlined above may be summarized as problems related to the geologic structure and tectonics, sedimentary facies and processes, and origin and development history of the seas around the Korean Peninsula. They include the epicontinental Yellow Sea and the shelf of the northern East China Sea, the East Sea (or western part of Sea of Japan) with narrow continental shelf and deep basins, and the South Sea with numerous rocky embayments.

It is important to describe briefly the geology of the Korean Peninsula before discussing its surrounding seas.

Korean Peninsula

Geologically, the Korean Peninsula forms part of the Sino-Korean Platform and is underlain by Precambrian metamorphic rocks. The southern half of the peninsula consists of a number of northeast–southwest trending tectonic units (fig. 1.4a): Precambrian Gyeonggi Massif, Ogcheon (Okchon) Fold Belt (? late Proterozoic to Cretaceous), Taebaegsan Basin (Paleozoic), Precambrian Yeongnam (Ryeongnam) or Sobaegsan Massif, and Cretaceous Gyeongsang (Kyungsang, Kyeongsang) Basin (S.M. Lee 1974; Reedman and Um 1975; M.S. Lee 1981). Tertiary sedimentary rocks occur locally in the southern coast (Pohang Basin) and in Jeju (Cheju) Island.

Gyeonggi and Yeongnam Massifs

The Precambrian Gyeonggi and Yeongnam massifs consist largely of granitic gneiss and schist and occur in the middle and southern parts of the peninsula separated by the Ogcheon Fold Belt (fig. 1.4b). The Ogcheon Fold Belt consists of a complex metasedimentary sequence of pelitic rocks. In large part, the Gyeonggi Massif includes banded biotite gneiss, quartzo-feldspathic gneiss, migmatitic granite gneiss, and augen gneiss. Mineralogically they are composed largely of microcline, plagioclase, biotite, and hornblende with accessory garnet, zircon, epidote, andalusite, sillimanite, and cordierite associated largely with the amphibolite facies (Na 1980). Various types of schist, such as biotite schist, biotite-sillimanite schist, graphitic schist, quartz-muscovite schist, and amphibolite and marble are also interspersed in the gneiss. Both the gneiss and schist are believed to have been derived from sedimentary sequences (Reedman and Um 1975). Also occurring in the extensive metasedimentary group are quartzite and quartz-sericite schist within the Gyeonggi Massif resting unconformably on the older basement (Son 1971a, 1971b; Na 1972; and others). The Yeongnam Massif consists commonly of banded biotite gneiss and augen gneiss with lenses of quartzite, quartz-sericite schist, biotite schist, amphibolite, and marble, much similar to those of the Gyeonggi Massif. Large bodies of anorthosite, gabbro, and diorite intrusions are present also. According to the isotopic age determinations on samples, the massif ranges in age from about 2800 to 800 Ma (fig. 1.4c), suggestive of various metamorphic events since Archaeozoic time (Hurley et al. 1973; Na and Lee 1973; Fullagar and Park 1975).

Ogcheon Fold Belt
Clastic rocks of the Ogcheon Group comprise original mudstones, sandstones, limestones, and pebbly sandstones and mudstones (M.S. Lee and Park 1965; Son 1970; O.J. Kim 1970; Reedman et al. 1973; and others). They crop out across the southern Korean Peninsula trending northeast-southwest (fig. 1.4b). Amphibolites and large bodies of granitic plutons occur also in the periphery of the basin. The sediments were deposited in an elongate and progressively deepening basin toward the long axis of the basin. The occurrence of marginal sandstone and mudstone bordering the rifted Precambrian basement is widespread and replaced progressively toward the center by deeper marine facies. Debris flow deposits in the basin contain the pebbles that are rather well rounded and were resedimented probably via fluviatile or nearshore environments. The deposits represent slope facies and deeper, of the progressively subsiding Ogcheon Basin which was bounded by quartzite and limestone of shallow water origin (Chough 1981b; Chough et al. 1981a).

The Ogcheon Group rocks were folded intensely and metamorphosed during the Daebo Orogeny (Jurassic) undergoing three phases of strong deformation (Reedman et al. 1973; P.C. Kang and Chi 1980). These resulted in complex foliations, lineations, and metamorphism of various facies from greenschist to amphibolite (H.S. Kim 1971; Reedman et al. 1973). In most cases, the original bedding features are obscured by foliation and lineation. The stratigraphic correlation and the determination of the geologic age of the Ogcheon Group rocks have been hampered by the lack of fossil occurrence and the strong deformation during the orogeny. The basin was probably initiated either in the late Proterozoic (O.J. Kim 1970; J.H. Lee 1972; Reedman and Um 1975) or in post-Cambrian (K.W. Kim and Lee 1965; Son 1970) continuing into Cretaceous as postorogenic sequences. General agreement has yet to be made on the lithostratigraphic units and their relationships in the entire Group. (See P.C. Kang and Chi 1980 and O.J. Kim and Yoon 1980 for a review.)

Taebaegsan Basin
In the northeastern margin of the Ogcheon Fold Belt is the Taebaegsan Basin (fig. 1.4a) that consists of the Gangweon Synthem (Joseon Group) (Cambro-Ordovician) and the Jangseong Synthem (Pyeongan Group) (Carboniferous to early Triassic) (fig. 1.4c) (K.H. Chang 1975; Cheong 1982). The former synthem consists mainly of limestones overlain unconformably by the latter which is primarily of nonmarine clastics including important coal measures. Both in the Ogcheon and Taebaegsan Basins are the Bansong (Jurassic) and Gyeongsang (Cretaceous) nonmarine, postdeformational sequences of subordinate order. A Silurian sequence occurs in the uppermost part of the Gangweon Synthem (Joseon Group) found recently by H.Y. Lee (1980).

Mesozoic Orogeny
A series of tectonic movements accompanied by granite intrusion, in Mesozoic time disturbed the pre-Mesozoic sequence in the southern peninsula (fig. 1.4b and 1.4c). The first disturbance (Songrim Disturbance) occurred in middle to late Triassic time mostly in the northern part of the Korean Peninsula. The Daedong Group rocks were deposited locally in the postdeformational inland basins. The second orogeny (Daebo Orogeny) occurred in the middle Jurassic and affected most of the middle part of the peninsula with extensive granite intrusion trending northeastward. The metamorphic grade increases from greenschist to amphibolite facies toward the northeastern and south-

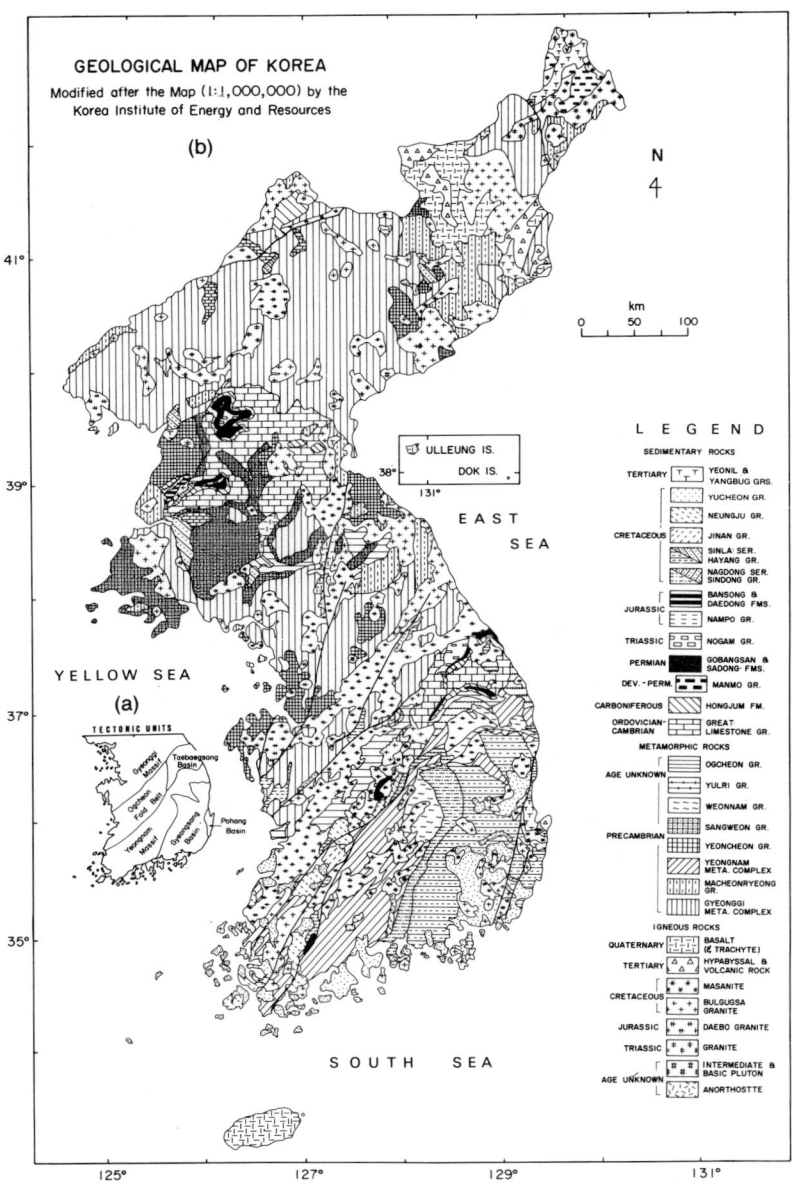

Figure 1.4. a: tectonic units; b: simplified geologic map of the Korean Peninsula; and c: the major stratigraphic units. b: after Korea Institute of Energy and Resources (1981a). c: after Reedman and Um (1975) and M.S. Lee (1981).

Figure 1.4c.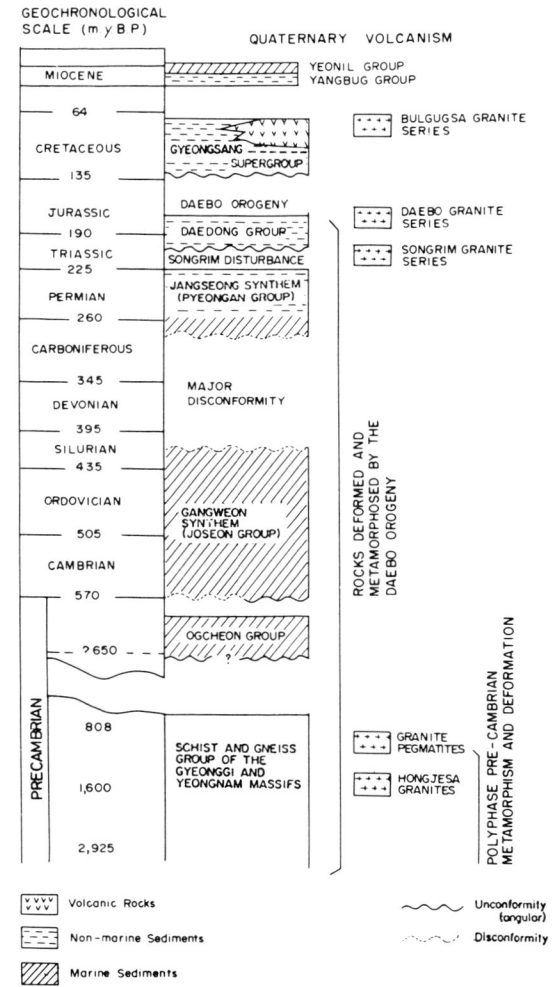

western margins of the Ogcheon Fold Belt (H.S. Kim 1971; Reedman et al. 1973). Mineral assemblages in pelitic rocks include almandite, staurolite, and kyanite suggestive of an intermediate pressure metamorphic environment.

Granitic rocks (granite, granodiorite, and tonalite) that were intruded in the peninsula during the Mesozoic are generally youthful southward. They range in K-Ar and Rb-Sr age from about 200 my (Triassic) in the northern part of the Korean Peninsula to about 160 my (Jurassic) mainly in the middle part close to the Ogcheon Fold Belt (fig. 1.4b). Granitoids that were intruded in Cretaceous to Tertiary time (120 to 50 Ma) tend to occur in many parts of the peninsula (M.S. Lee, in press). In the Gyeongsang Basin it intruded about 100 my ago associated with volcanic activity. Workman (1972) and B.K. Park and Do (1973) ascribed the southward youthful magmatism to a fixed thermal plume (hot spot) above which the peninsula moved clockwise north- and northeast-ward. The southward youthful magmatism as well as metamorphic facies led S.M. Lee (1974) to propose an arc-trench system in which a subducting oceanic plate retreats oceanward with time.

Gyeongsang Basin
The Gyeongsang Supergroup was formed in early Cretaceous in the southeastern part of the peninsula (fig. 1.4a and 1.4b) in postorogenic inland fluviolacustrine basins (K.H. Chang 1977). The clastics were derived from the northwest (K.H. Chang and Kim 1968) resulting in a lateral facies change from alluvial fan to flood plain to the southeast (H.I. Choi 1981). Intermediate and acidic volcanic rocks followed by granitic intrusions in late Cretaceous occur in the basin and mineral occurrences of copper, lead-zinc, and fluorite-tungsten-molybdenum deposits are associated with them. According to Sillitoe (1977), this mineralization pattern is suggestive of shallow subduction beneath Korea by the Pacific Plate during late Cretaceous. The granites and the associated rocks are related probably to subduction (Jin 1981).

Cenozoic Events
During the Tertiary period, a relatively shallow marine basin (Pohang Basin) was formed on the southeastern corner of the

peninsula (fig. 1.4a and 1.4b) in which clastics were deposited. These clastics were associated with volcanic activity extending into most of the present continental shelf and the nearby area (Yeonil Group). Also nonmarine clastics were deposited locally (Yangbug Group). Quaternary volcanism, composed mainly of alkali basalts, was active in the islands of Jeju and Ulleung.

2 Epicontinental Sea (Yellow Sea)

Physiography

The epicontinental sea which includes the Yellow Sea (or West Sea by Koreans) is about 500,000 km² in area, and is arbitrarily bordered by the northern East China Sea by a line connecting Jeju (Cheju) Island and south of the Yangtze River, and the continental shelf southeast of it (fig. 2.1). The shallow sea area south of the peninsula between Jeju Island and Tsushima Island has been named the South Sea by Koreans.

The Yellow Sea is characterized by a flat, broad, and featureless sea floor with average water depth of about 55 m (maximum less than about 100 m, fig. 2.1). The western part of the sea floor is bordered by the deltas of both the Hwangho and Yangtze rivers and the isobaths are parallel approximately to the coastline. The eastern Yellow Sea (fig. 2.2) is fringed by numerous islands and characterized by a long stretch of tidal flat along the coast. Tidal sand ridges are ubiquitous in the eastern Yellow Sea in water depth less than about 70 m, trending slightly oblique to the coastline (fig. 2.2). The sea floor deepens progressively toward the axis that lies in about the eastern two-thirds of the sea (fig. 2.1). The sea floor of the shelf deepens progressively southeastward to form the northern extension of the Okinawa Trough (fig. 2.3). Rock outcrops on the sea floor are only known locally to the west of the Korean Peninsula, the short continuation of Jeju Island and at Socotra rock southwest of Jeju Island. The sea floor around Jeju Island is deeper, exceeding about 100 m.

Geologic Structure

Acoustic Basement
The central part of the Yellow Sea consists of thick Tertiary sediment (fig. 2.4a) overlying acoustic basement of igneous, metamorphic, and sedimentary rocks that crop out in the surrounding landmass (Emery et al. 1969). The sediment sequence is thicker (or the basement is deeper) toward China and is bounded on the western margin by a probable fault (C.S. Kim et al. 1969; Wageman et al. 1970). The shoaling of the acoustic basement toward the Shantung Peninsula is probably due to an extension

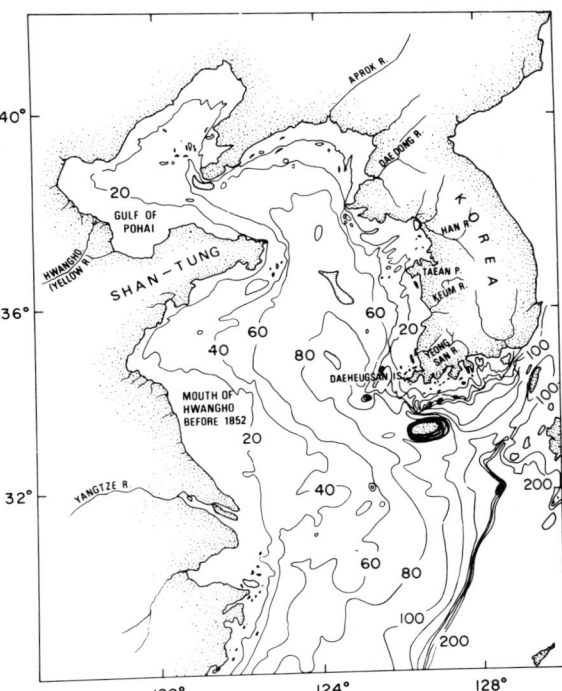

Figure 2.1. Index map showing the geographic names adjacent to the Yellow Sea. Contours in meters. After Chough and Kim (1981) by permission of the Journal of Sedimentary Petrology.

Epicontinental Sea (Yellow Sea)   17

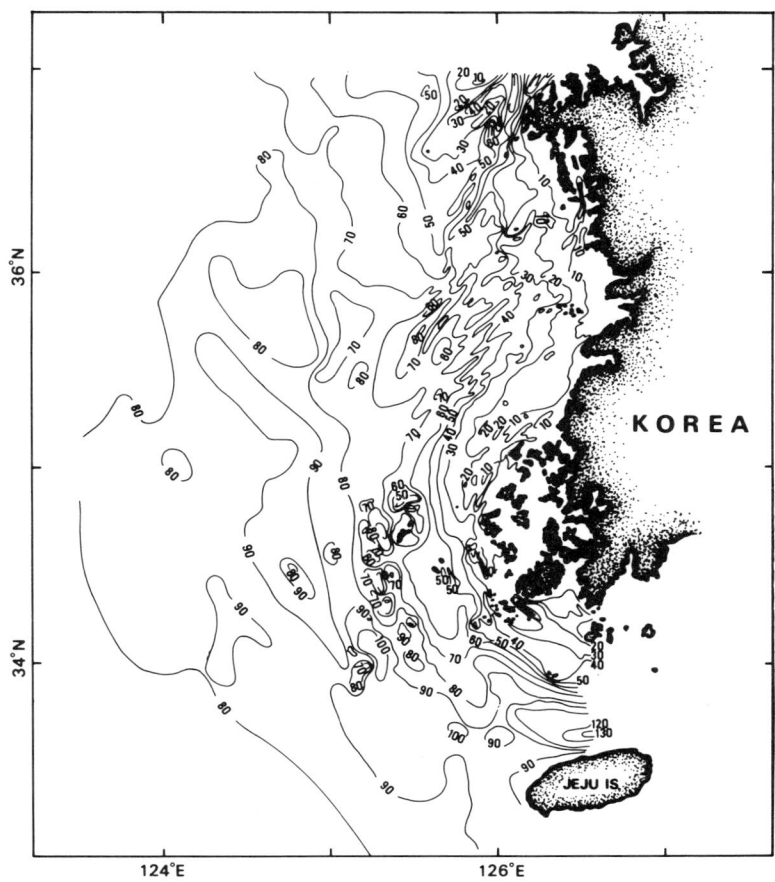

Figure 2.2. Detailed bathymetric chart of the southeastern Yellow Sea. Contours in meters. After C.S. Kim et al. (1982) courtesy of the Korea Institute of Energy and Resources Bulletin.

18 Marine Geology of Korean Seas

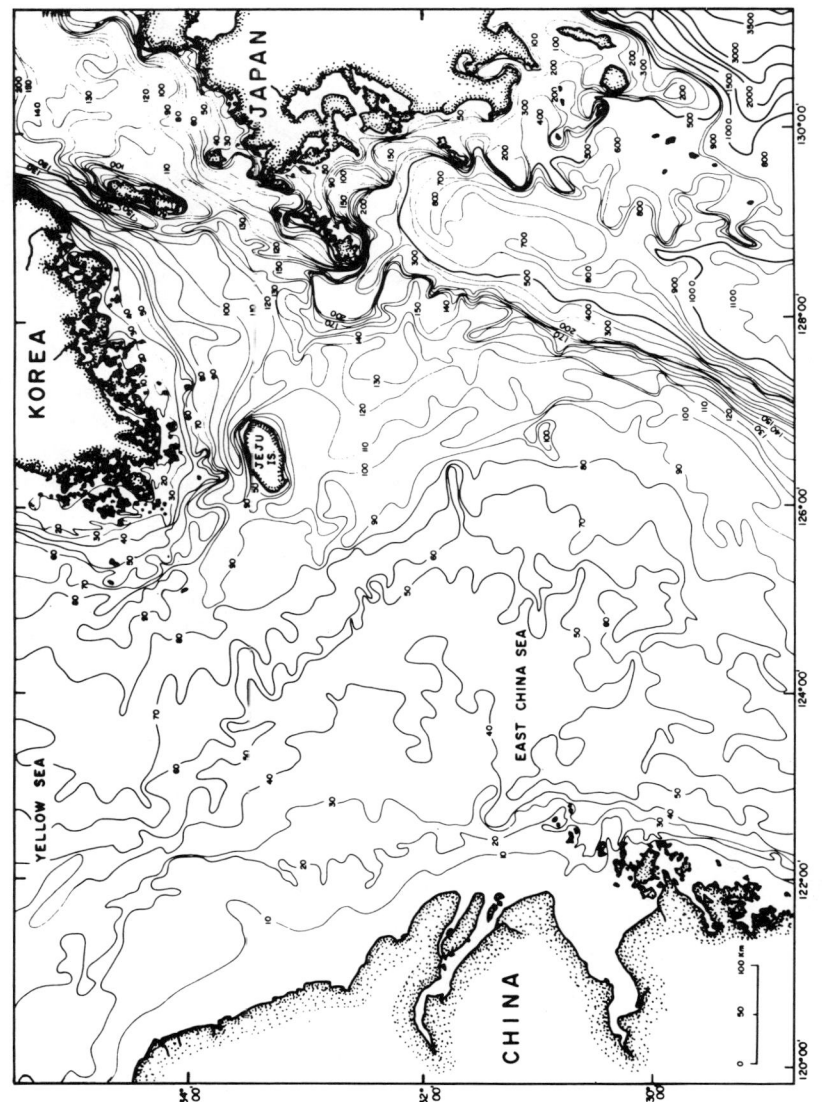

Figure 2.3. Detailed bathymetric chart of the southern Yellow Sea and the northern East China Sea. Contours in meters. Modified after Hahn (1979) courtesy of the Korea Ocean Research and Development Institue Bulletin.

of the Shantung-Laoyehling Massif that is continuous northeasterly to the northern Korean Peninsula. Toward the southern Yellow Sea and the South Sea the basement also shoals (figs. 2.4a, 2.4b, and 2.5). This shoaling ridge across the entrance of the Yellow Sea appears to be related to the Fukien Massif in China. The Fukien Massif is most likely continuous to the Yeongnam (Reinan, Ryongnam; Sobaegsan) Massif in the southern Korean Peninsula (fig. 2.6) which was uplifted in the Mesozoic. The magnetic profiles are slightly undulatory where the basement is deep, but are irregular where the basement is shallow (figs. 2.4a and 2.4b) (Koo et al. 1970; Wageman et al. 1970; Bosum et al. 1971). The basement shoal near the southwestern tip of the Korean Peninsula, the Heugsan (Huksan) Platform, is probably a seaward extension of the Ogcheon Fold Belt and Yeongnam Massif (fig. 2.6), and is covered with a thin veneer of deformed Paleogene and unconformably overlying Neogene sediments (fig. 2.7). Depth to the basement on the platform is less than about 500 m.

The Jeju volcanic belt (fig. 2.6) (Frazier et al. 1976), whose existence is also evident in magnetic profiles (Koo et al. 1970; Bosum et al. 1971; K.J. Cho 1979), extends westward from Jeju Island, and is composed probably of intrusive rocks formed during the late Tertiary volcanism. High magnetic anomalies generally trending northeast–southwest (fig. 2.5) suggest that shallow basement or younger volcanic rocks underlie the sediments. Local high anomalies trending northwest–southeast have also been identified in the west of the Jeju Island (fig. 2.5) (Bosum et al. 1971; K.P. Park 1982). In the South Sea, the strong anomaly is probably due to Cretaceous volcanics. The depth to the igneous basement ranges from 1.6 to 3.3 km in this region (Bosum et al. 1971). In the Korea Strait, the anomaly is weak, suggestive of nonmagnetic basement (K.J. Cho 1979). Other shoals such as Socotra rock that crop out southwest of Jeju Island also consist of Tertiary volcanic rocks.

In the shelf area, an intensely folded thick sedimentary sequence trends northeast–southwest (fig. 2.4b), underlain partly by intrusive rocks that are exposed near western Kyushu (Wageman et al. 1970). Here, aeromagnetic data of Koo et al. (1970) and Bosum et al. (1971) also confirm the possible existence of a thick sedimentary sequence. This is the Taiwan-Sinzi Folded

20   Marine Geology of Korean Seas

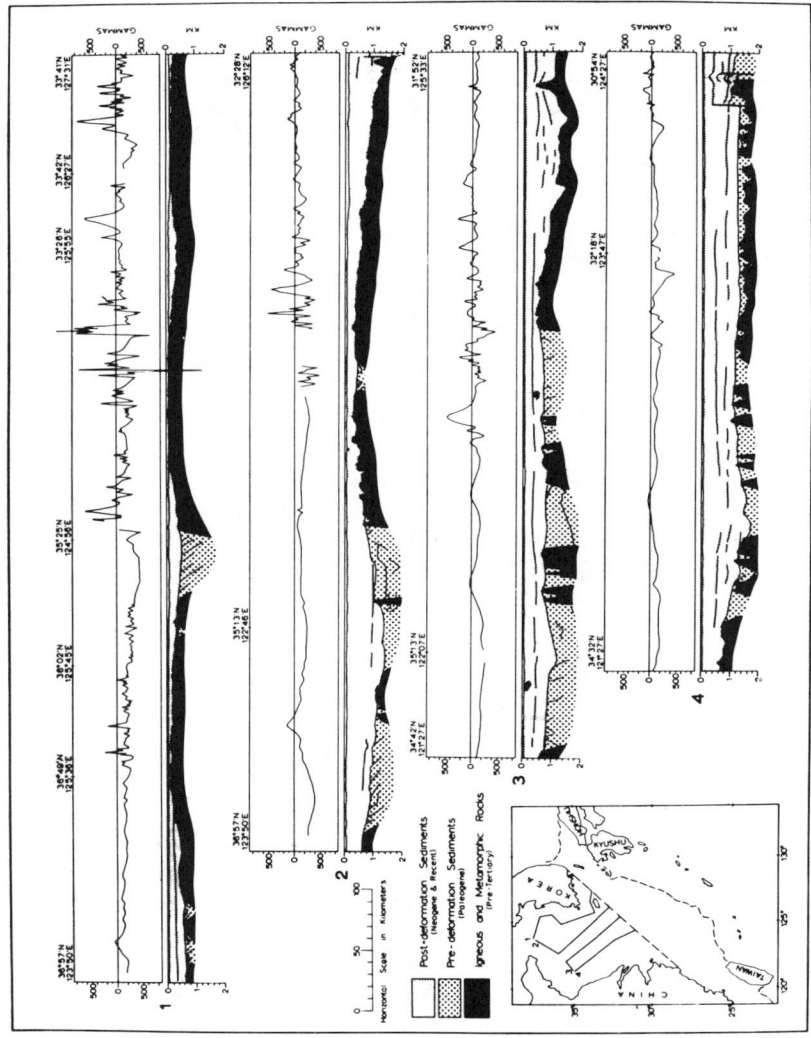

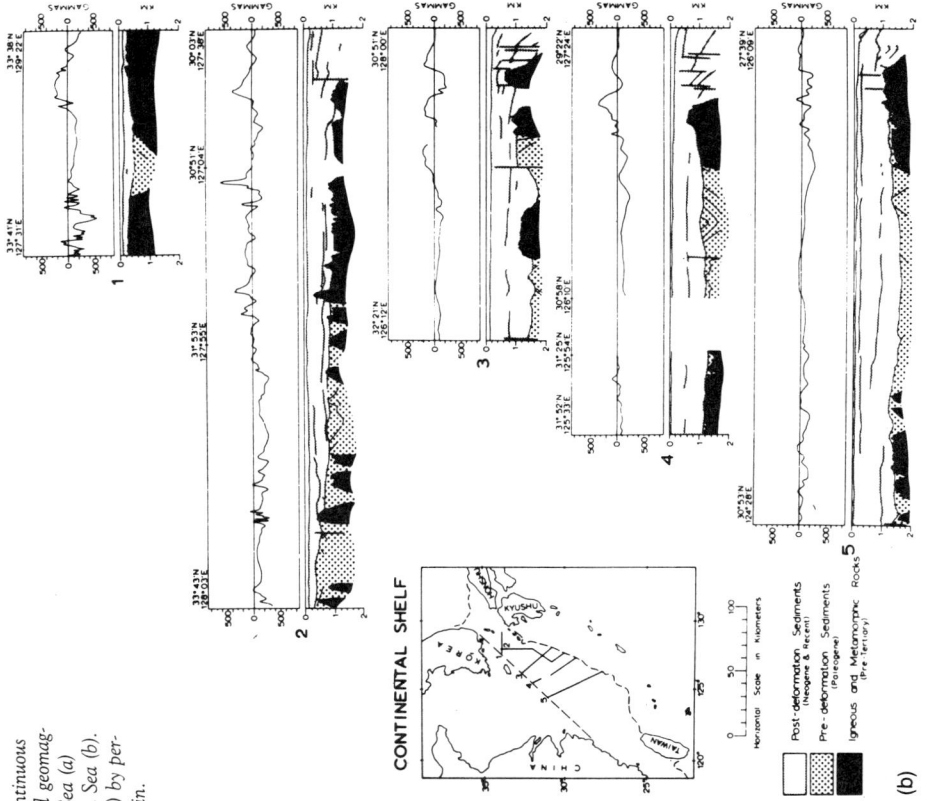

Figure 2.4. Interpreted continuous seismic reflection profiles and geomagnetic profiles of the Yellow Sea (a) and the northern East China Sea (b). After Wageman et al. (1970) by permission of the AAPG Bulletin.

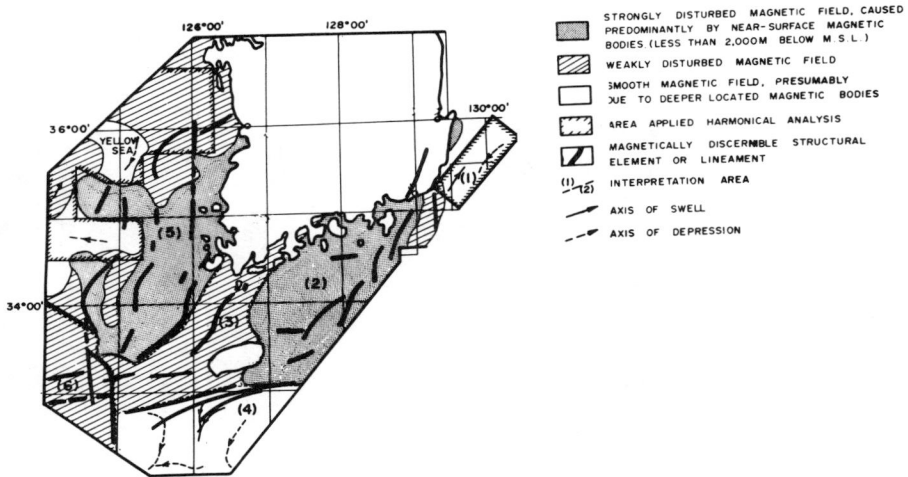

Figure 2.5. Structural map based on geomagnetic anomalies in the southeastern Yellow and South seas. After Koo et al. (1970) and Bosum et al. (1971) by permission of the CCOP Technical Bulletin.

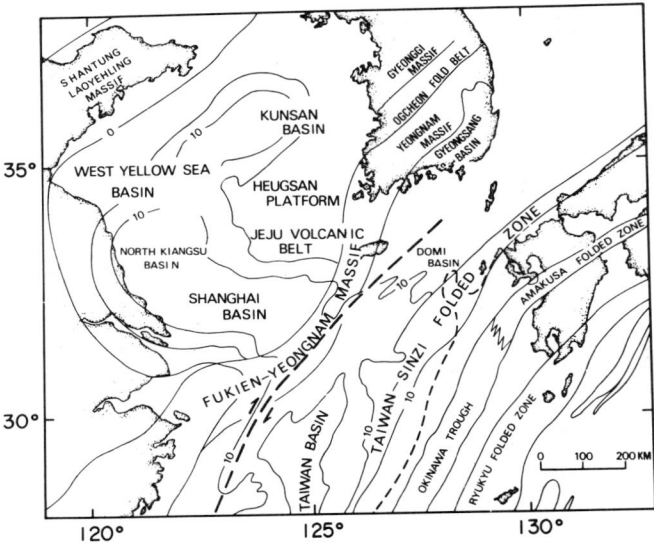

Figure 2.6. Simplified map of tectonic units in the Yellow and East China seas correlated with those on the adjacent landmass of China, Japan, and Korea. Contours indicate sediment thickness in hundreds of meters. Modified after Emery et al. (1969) by permission of the CCOP Technical Bulletin.

Epicontinental Sea (Yellow Sea) 23

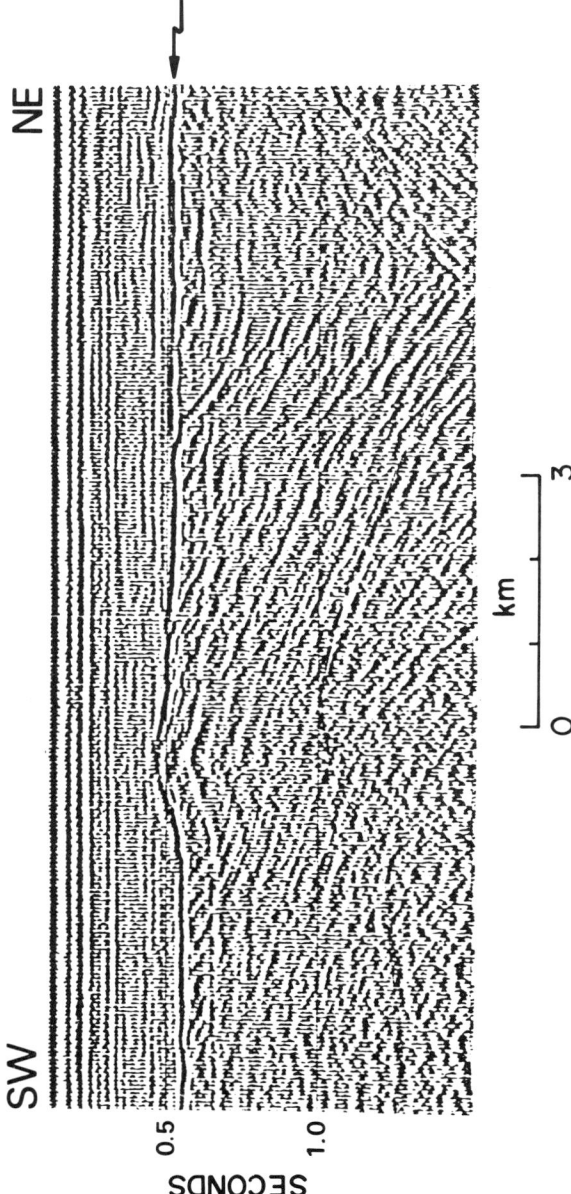

Figure 2.7. Seismic reflection profile in the Yellow Sea on the Heugsan Platform near Soheugsan Island (33°35'N, 124°07'E) showing the regional unconformity (arrow) of Plio-Pleistocene age overlying the deformed Tertiary sequence. Vertical scale in two-way traveltime (seconds). Profile courtesy of the Korea Institute of Energy and Resources.

Zone (fig. 2.6) whose deformation occurred probably during Neogene time (Emery and Niino 1967; Emery et al. 1969). The Neogene sequence in Taiwan, a southwestern extension of the Taiwan-Sinzi Folded Zone, is more than 5000 m thick (Meng 1968). In the northeast the sequence extends near Kyushu, Japan (Emery and Niino 1967).

Sedimentary Sequence
The sedimentary sequence in the Yellow Sea overlying the acoustic basement is more than 2000 m thick and consists of two units (Emery et al. 1969; Frazier et al. 1976) (fig. 2.7): 1) immediately overlying the basement are deformed sedimentary strata of Paleogene age and earlier; and 2) the overlying Neogene undeformed strata are bound by a major regional unconformity following Miocene orogeny. The sedimentary strata are thicker and flat-lying to the southwest and appear to be deformed more near the Korean Peninsula where numerous dipping beds and synclinal structures are present (Wageman et al. 1970). The regional unconformity becomes progressively shallow toward the west coast of the Korean Peninsula (fig. 2.8), surfacing near the Heugsan Platform. The platform is covered with thin Neogene marine sediments, suggestive of its submergence since Pliocene time. The sediment isopach maps, compiled by Wageman et al. (1970), both of the total sedimentary column overlying the acoustic basement (fig. 2.9a) and of the postdeformational (Neogene) sequence (fig. 2.9b) overlying the major unconformity, reveal that the depositional basins have generally been shifted west- and southwest-ward in Neogene time.

The Western Subbasin is characterized by a small but deep depression between the West Yellow Sea Basin and Kunsan Basin (figs. 1.3 and 2.6), formed by a fault with deposition of over 5000 m of sediment (Frazier et al. 1976). As revealed in a well (IIH-1xA) drilled in the basin (fig. 1.3), the postdeformational sequence (719 m thick) is composed of semiconsolidated gray sandstone and mudstone with intercalated lignite (Frazier et al. 1976). The predeformational sequence consists mainly of reddish brown mudstone, marlstone, sandstone, and conglomerate (719–3284 m). Below this depth (3284–3467 m), mostly thin-bedded reddish brown silty shale occurs, whose matrix is oxidized and iron-stained. According to pollen study, the upper-

Epicontinental Sea (Yellow Sea) 25

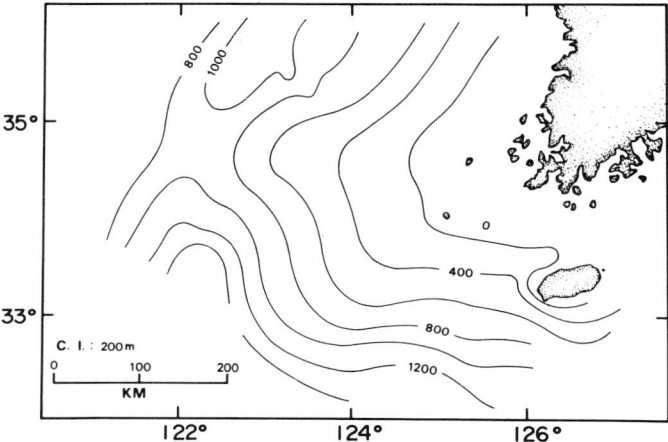

Figure 2.8. *Subsurface depth of the regional unconformity in the southeastern Yellow Sea. After Frazier et al. (1976) by permission of AAPG Bulletin.*

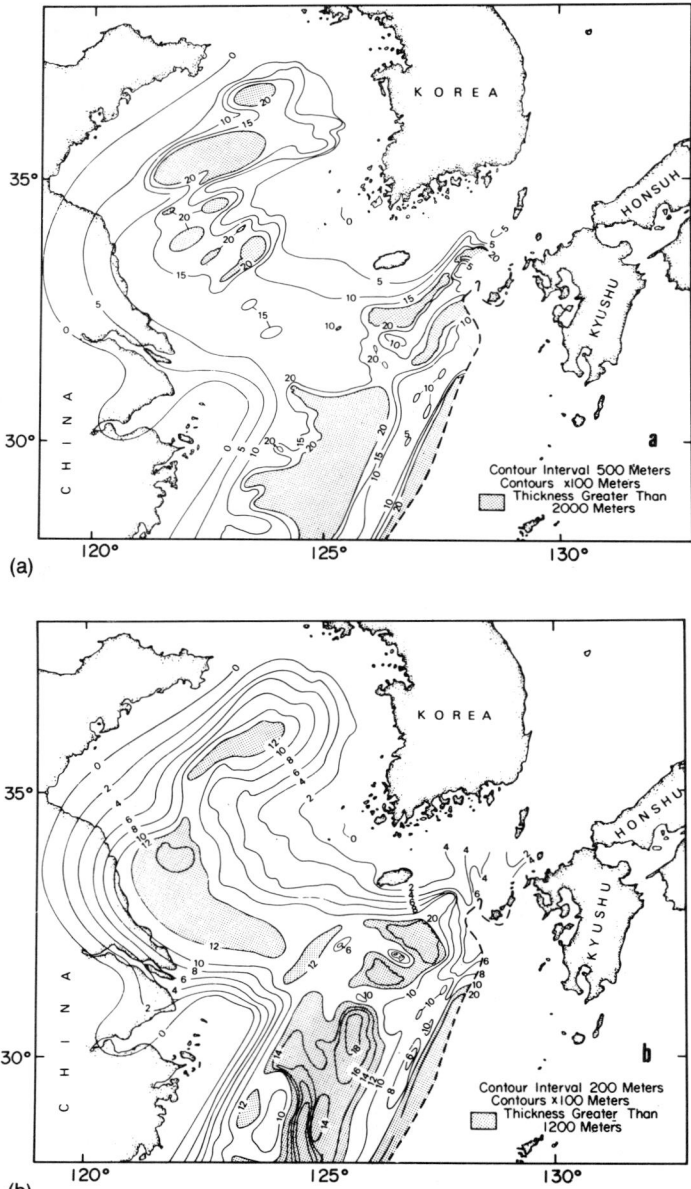

Figure 2.9. Isopach map of total (a) and postdeformational (b) sedimentary sequences in the Yellow and East China seas. After Wageman et al. (1970) by permission of AAPG Bulletin.

most unit (311–497 m) was assigned to the Holocene; 549–716 m, Oligocene to Holocene; 780–3230 m, Tertiary; 3261–3467 m, unknown. The presence of charophytes and freshwater ostracods and absence of nannofossils in the predeformational sequence indicate that it was deposited in continental environments such as alluvial fan, fluviatile, and flood plain (Frazier et al. 1976). The seismic velocity through the predeformational sequence is high, ranging from 3050 m/s to 4575 m/s. This is probably due to a high carbonate content in calcareous shale and claystone (Frazier et al. 1976).

Sediments in the Kunsan (Gunsan) Basin also consist of thick (ca 3000 m) predeformational sequence and thin (ca 600 m) postdeformational sediment overlying the regional unconformity, observed in an exploratory well (IIC-1x) in the western flank of the basin (fig. 1.3). Other sedimentary basins (fig. 2.6) identified in the Yellow Sea include the West Yellow Sea Basin, North Kiangsu Basin, and Shanghai Basin, all parts of one inland depression probably separated from each other during the early stage (Paleogene) of basin development. A small Socotra Subbasin east of Shanghai Basin also consists of thick (more than 3600 m) Tertiary sedimentary sequence.

In the South Sea, Sora and Domi basins are outlined in the east of Jeju Island where two wells (Sora-1 and Domi-1) (fig. 1.3) penetrated into more than 3000 m of Neogene sediments. Here sediments were deposited probably in alluvial fan to coastal plain environments. In the shelf area further south, the predeformational sequence is concentrated in the northern part toward the epicontinental sea (e.g., Fukue Basin) and underlain by an angular unconformity that is about 500–1500 m deep (Wageman et al. 1970). The postdeformational sequence is also composed of Neogene sediments ranging in thickness from tens of meters off Jeju and China to more than 2000 m southeastward away from the continent (see Addendum).

Shallow Structure

Shallow geological structure of the west coast offshore of the Korean Peninsula east of 125°E has been revealed through a series of seismic (both air gun and Uniboom) and magnetic surveys by the KIER teams (fig. 1.2) (Koo et al. 1971; S.J. Yang et al. 1971; C.S. Kim et al. 1972; Koo 1972; J.H. Lee et al. 1972; S.W. Kim et al. 1980a, 1980b; C.M. Kim et al. 1981; S.W. Kim and Min 1981; C.S. Kim et al. 1982; and others), whose results have recently been published as a map series. Recently, a 3.5 kHz subbottom profiler was used to obtain high resolution near-bottom profiles.

Acoustic Basement
The acoustic basement appears to be covered with a thin veneer of unconsolidated or semiconsolidated Quaternary sediments less than about 30 m thick nearshore (locally up to 60 m) and slightly thicker seaward (fig. 2.10). On the Heugsan Platform, sediments are thin and often basement rocks are exposed on the sea floor (fig. 2.11). This occurs near islands where strong tidal currents preclude the deposition of sediments. The acoustic basement consists of the seaward extension of on-land rocks such as Precambrian gneiss and schist, Jurassic granite, metasedimentary rocks, and others (C.J. Cho and Choi 1970; N.Y. Park et al. 1972) that can be correlated positively with the weak magnetic intensity (fig. 2.11). Locally, some extrusive rocks and schists with magnetite bodies signal strong magnetic intensity (S.J. Yang et al. 1971; C.M. Kim and Lee 1974a; C.M. Kim et al. 1981).

Sedimentary Sequence
The sediments can be divided into several layers (or sequences) by a number of mid-reflectors (figs. 2.11 and 2.12). The reflectors are named tentatively here α, β, and γ, etc. from the top. Each consists of probably hardground corresponding to the sea-level lowstand during the Quaternary period. The first reflector, reflector α, resulted most likely from the Wisconsinan sea-level lowstand and occurs at various depths below the sea floor or in some areas exposed on the surface (figs. 2.12 and 2.13a). The sequence above the reflector α, designated sequence A (transgressive sequence), consists of unconsolidated probable Holocene sediments that were deposited during the postglacial transgression and ranges in thickness from a few meters to about 40 m (average thickness, 10 m) (figs. 2.13b and 2.14a).

Sequence B, designated as the sediment layer between the reflector α and the reflector β, also is distributed uniformly in

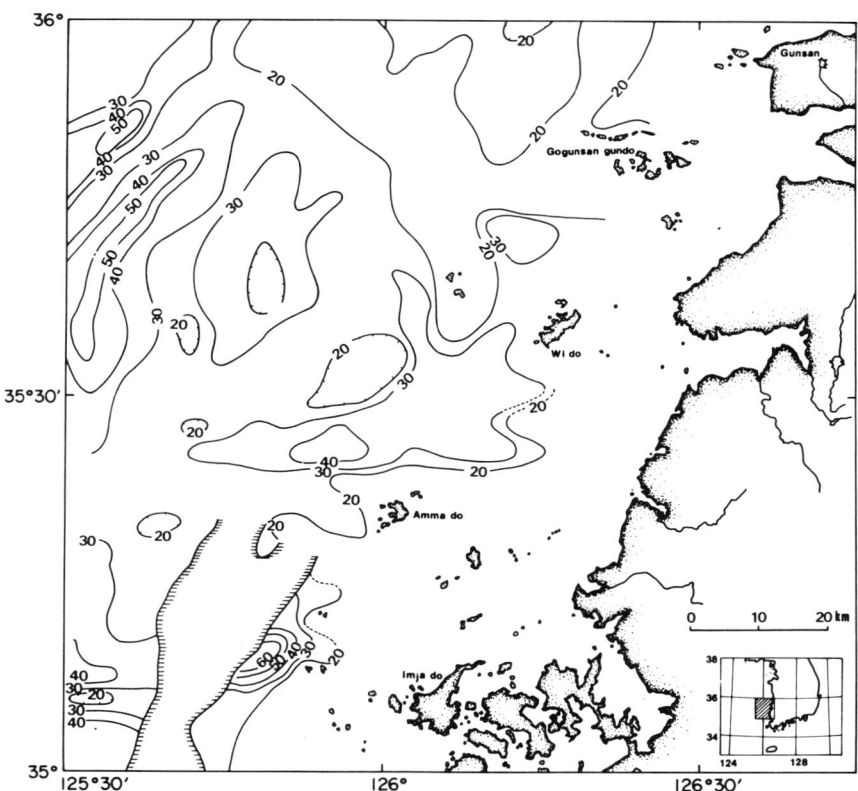

Figure 2.10. *Isopach map of sediment thickness above the acoustic basement (? Paleogene) in the southeastern Yellow Sea (see inset). Hatched area for gas-charged sediments. Contours in meters. After KIER Map Series (II) (1981b).*

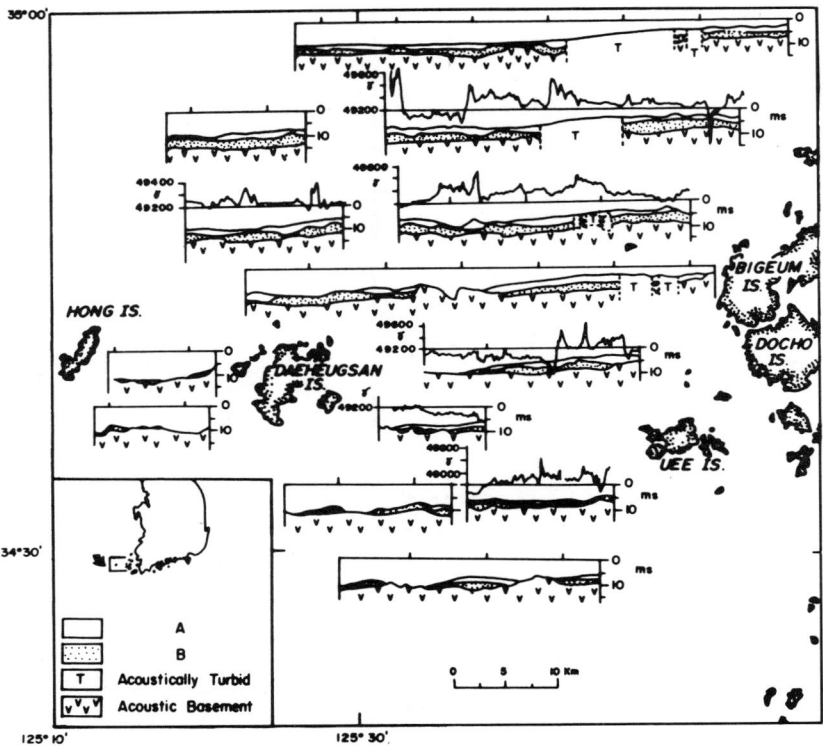

Figure 2.11. *Simplified shallow structure of the southeastern Yellow Sea (see inset) showing the surface sediments (sequences A and B), gas-charged acoustically turbid sediments (T), and the acoustic basement. Also shown is magnetic intensity in gamma (γ). Vertical scale in two-way traveltime in milliseconds. Modified after S.W. Kim et al. (1980a) courtesy of the Korea Institute of Geosciences and Mineral Resources Bulletin.*

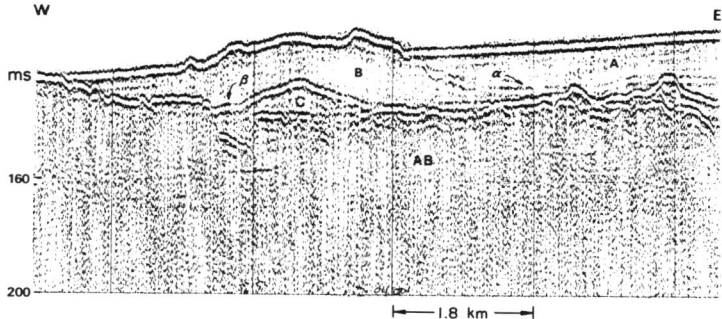

Figure 2.12. Uniboom profile showing the mid-reflectors α and β near the Hatae Island (34°20'N, 125°30'E). Also shown are the unconsolidated transgressive sequence (A) overlying α, sequence B (regressive sequence) below it and the sequence C above the acoustic basement (AB). Vertical scale in two-way traveltime in milliseconds. For location of the profile see figure 2.14a. Profile courtesy of the Korea Institute of Energy and Resources.

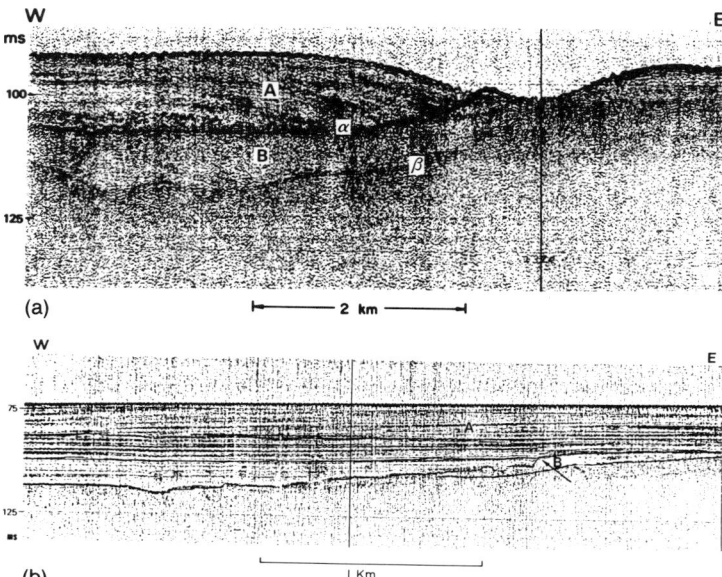

Figure 2.13. a: 3.5 kHz seismic reflection profile (ORE) east of Hatae Island (34°22'N, 125°30'E) showing inclined strata in sequence A (transgressive sequence) and underlying mid-reflector α, which is exposed on the sea floor toward the right. Sequence B and the underlying mid-reflector β are also shown. b: Transgressive sequence A showing a coastal onlap (arrow) toward the Korean Peninsula. Vertical scale in two-way traveltime in milliseconds. For location of the profiles see figure 2.14a. Profiles courtesy of the Korea Institute of Energy and Resources.

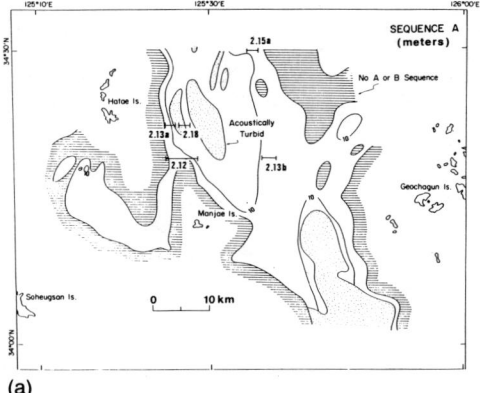

(a)

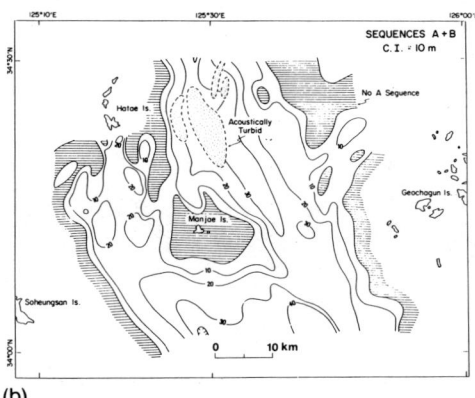

(b)

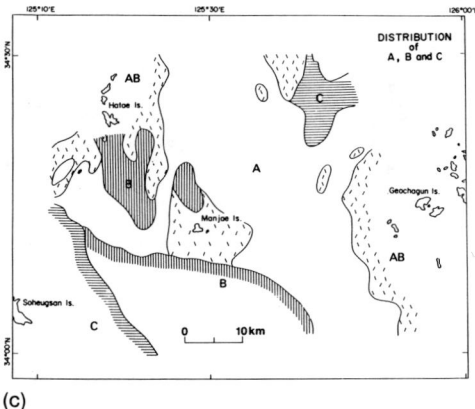

(c)

Figure 2.14. *a: Isopach map of sequence A; b: Isopach map of sequences A and B; c: Surface distribution of sequences A, B, C, and acoustic basement (AB) in the southeastern Yellow Sea northeast of Soheugsan Island. Profile courtesy of the Korea Institute of Energy and Resources.*

most of the nearshore area of the southeastern Yellow Sea with a thickness of up to 50 m (fig. 2.14b) (S.W. Kim et al. 1980a; C.S. Kim et al. 1982). Sequence B reveals complicated structure, that is, beds are often inclined, incised in many areas and refilled with sediments (figs. 2.13a and 2.13b). Sequence B represents most likely the sediments deposited during the Wisconsinan glacial regression (regression sequence). Sediments have also undergone erosion and the underlying beds are truncated where the reflectors are exposed (figs. 2.12 and 2.13a). Layer C, the sediment layer between the reflector β and the acoustic basement, occurs locally and is exposed on the sea floor adjacent to the islands (fig. 2.14c) where it is scoured by strong tidal current.

Reflector α is diachronous in that it is the datum plane on which postglacial deposition has occurred. Dated in the Gamagyang Bay (H.J. Kang and Chough 1982) it formed prior to about 4500 to 5000 y B.P. On the other hand, this datum plane was probably exposed, extrapolated according to the sea-level curve of Emery et al. (1971), until about 9000 y B.P. further offshore. High resolution seismic profiles also reveal numerous paleochannels that are approximately perpendicular or subparallel to the shoreline. These channels are most likely erosional channels that were activated during the lower stand of sea level during the last glacial period.

The phenomenon of nondeposition or erosion of postglacial sediments is also encountered where strong currents meet a topographic high or an island. Here, sediments are often stripped off forming moats and deposited elsewhere due to the strong currents, probably tidal currents in this case. This is also indicated by the sand dunes and large-scale tidal ridges (figs. 2.15a and 2.15b) (C.M. Kim and Lee 1974b; S.W. Kim et al. 1980a; C.M. Kim et al. 1981; C.S. Kim et al. 1982; Klein et al. 1982). Tidal currents measured in the area off Anmyeon Island range up to 2 knots flowing mainly northeast- and southwestward (fig. 2.15b). Sand ridges are about 1–3 m in height and spaced at about 90–155 m trending dominantly in the current directions.

In the South Sea north of Jeju Island, a thin sediment layer (less than about 30 m) occurs above the acoustic basement of probable Neogene semiconsolidated sedimentary sequence (figs. 2.16a and 2.16b). It is slightly thicker in the southern part (fig.

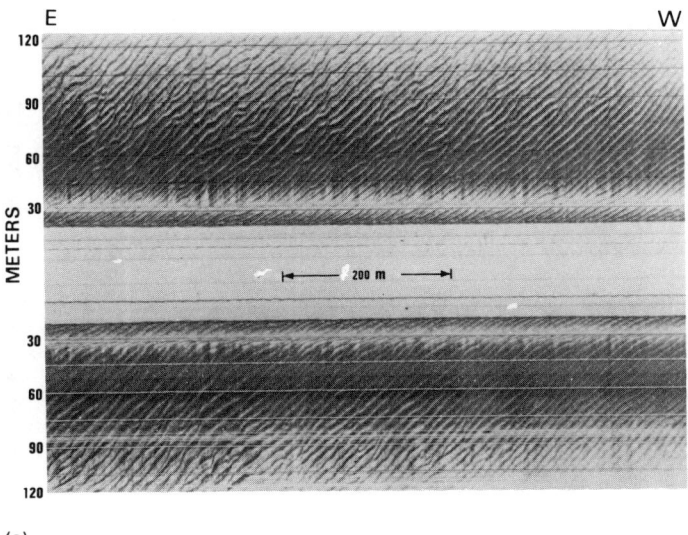

(a)

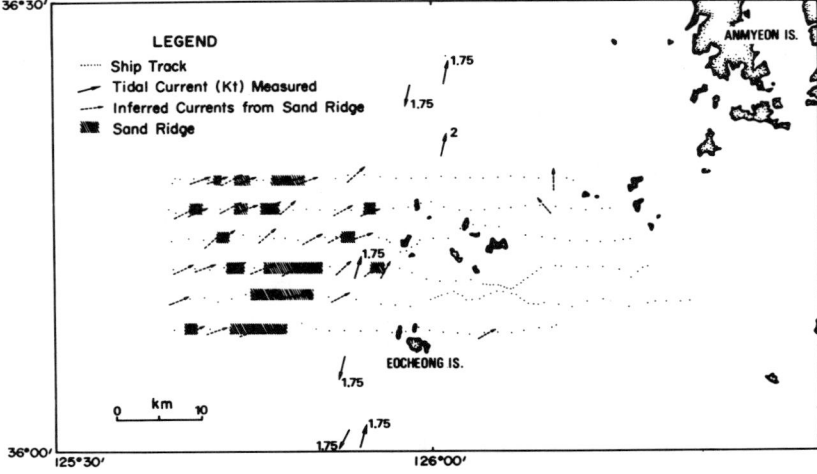

(b)

Figure 2.15. *a*: Side-scan sonar record over the sand dune field northeast Hatae Island (34°30'N,125°35'E). For location of the profile see figure 2.14a. Actual dimension of the asymmetrical dunes: length, about 3–3.5 m; height, 0.22–0.36 m; slope angle, about 25-30°. Currents from northeast to southwest. Courtesy of S.W. Kim, Korea Institute of Energy and Resources. *b*: tidal sand ridges found offshore of Anmyeon Island. After C.M. Kim and Lee (1974b) courtesy of the Geological Survey of Korea Report.

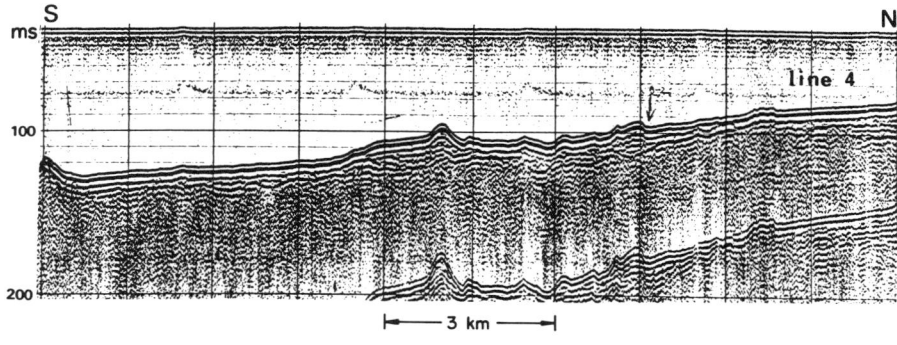

(a)

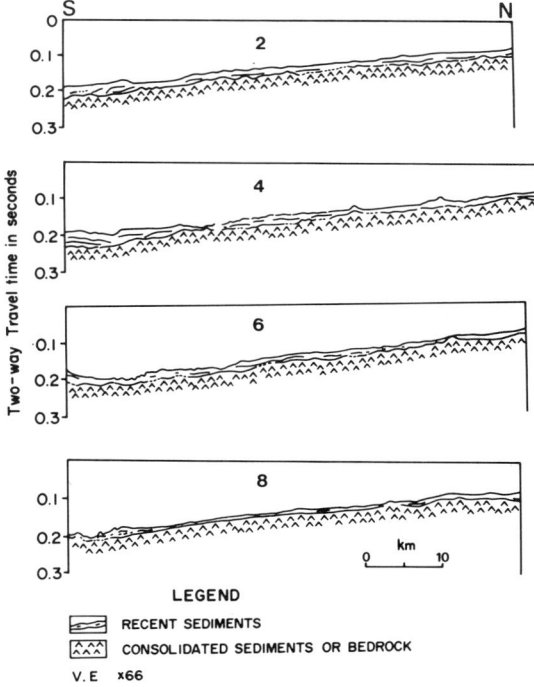

Figure 2.16. Uniboom profile (a) and line drawings (b) in the north of Jeju Island showing thin sediment layer of strong reflection and acoustic basement of probable Paleogene and thin Neogene sedimentary rock. Vertical scale in two-way traveltime in milliseconds. For location of the profiles see figure 2.17. After S.W. Kim et al. (1980b) courtesy of the Korea Institute of Geosciences and Mineral Resources Report.

(b)

2.17). The lack of sediment here is due to strong bottom or tidal current activity, through winnowing, removing the fine-grained material. This part of the sea is also characterized by coarse-grained sediment that contains abundant shell fragments resulting in a $CaCO_3$ content of up to 40% (S.W. Kim et al. 1980b).

Some parts of the sediment sequence on the eastern Yellow Sea are filled with probable gas (CO_2 and CH_4) resulting in an attenuation of acoustic wave energy (fig. 2.18). This feature is rather common along the southwestern coastal area (figs. 2.10 and 2.11). On the high resolution seismic profiles, some of the acoustically turbid layers show oblique and discontinuous stratification, suggesting that the acoustic turbidity could also result from the concentration of shells or other materials of different acoustic impedance.

Surface Sediments

Distribution

The epicontinental Yellow Sea is tectonically a stable, postglacially submerged depocenter of clastics derived from the adjoining landmass of China and Korea. On the northwest, the Hwangho River discharges large volumes (1080 million tons annually) of muddy sediments composed mainly of loess into the sea (Milliman and Meade 1983), dominating the sea with a pale yellow brown color (fig. 2.19a). The Yangtze River also empties large volumes of clastic materials (478 million tons annually) into the southwestern Yellow Sea and East China Sea (Milliman and Meade 1983) forming a belt of muddy sand and sandy mud in the southwestern Yellow Sea and northern East China Sea (fig. 2.19b). A number of short and steep Korean rivers discharge sandy sediments in relatively small amounts into the eastern Yellow Sea (Chough and Kim 1981). Sandy sediments are also dominant in the outer part of the shelf (fig. 2.19b) in response to winnowing of finer sediments by the Kuroshio (Niino and Emery 1961).

Detailed surveys made by the KIER mapping programs in the southeastern Yellow Sea also reveal the existence of substantial amounts of fine-grained sediments close to the peninsula (fig.

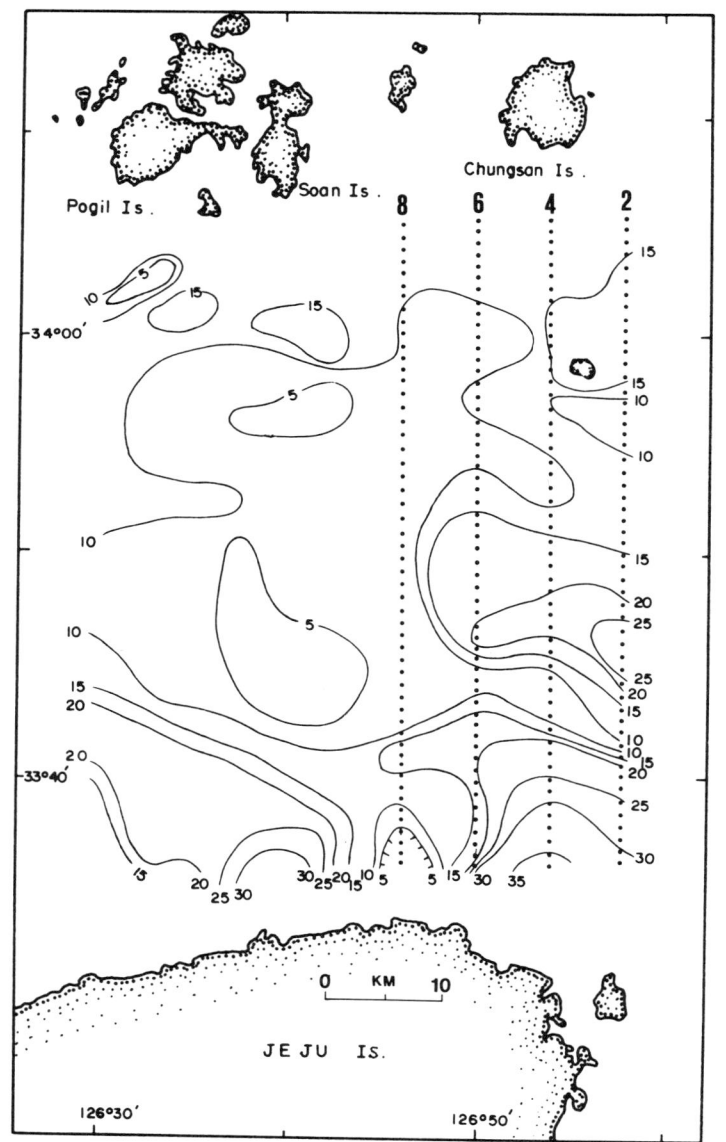

Figure 2.17. *Thickness (in meters) of Neogene sediment north of Jeju Island. After S.W. Kim et al. (1980b) courtesy of the Korea Institute of Geosciences and Mineral Resources Report.*

38 Marine Geology of Korean Seas

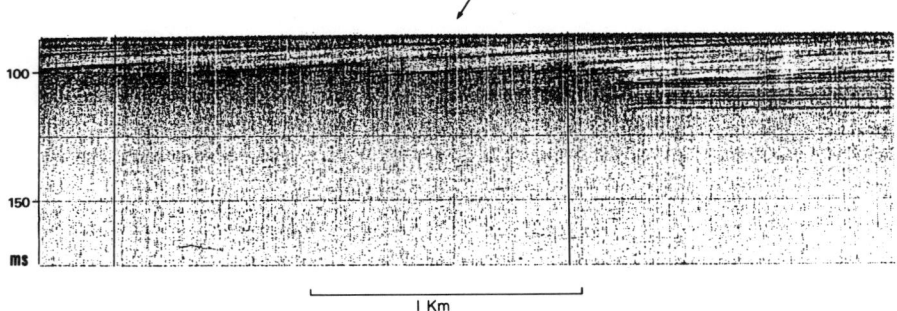

Figure 2.18. *3.5 kHz seismic reflection profile (ORE) in the east of Hatae Island (for location see figure 2.14a) showing the acoustically turbid layer below which acoustic energy attenuates. Vertical scale in two-way traveltime in milliseconds. Profile courtesy of the Korea Institute of Energy and Resources.*

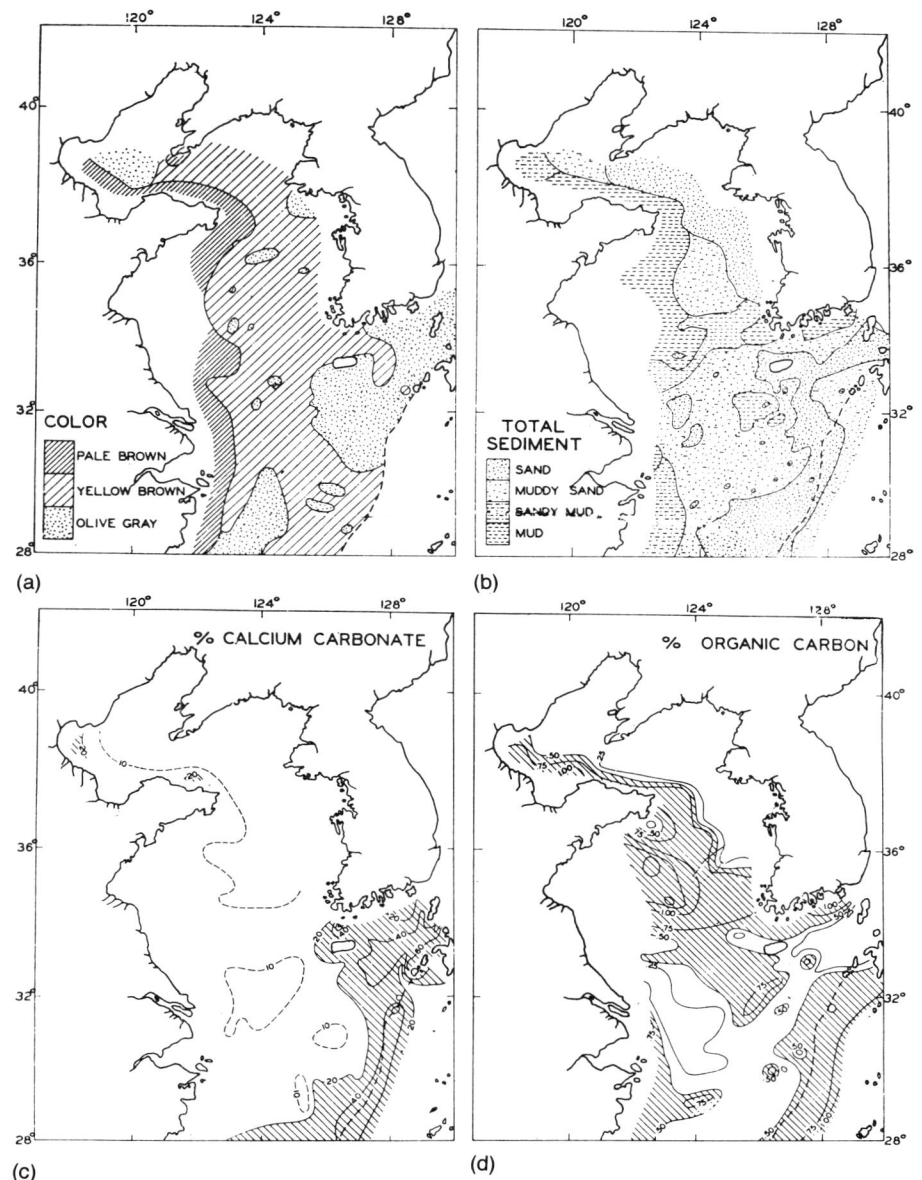

Figure 2.19. *General description of surface sediments in the Yellow and northern East China seas (after Niino and Emery, 1961). a: color; b: classification according to grain size; c: calcium carbonate (%), and d: organic carbon (%). Reprinted by permission of the Geological Society of America Bulletin.*

2.20) (N.J. Kim et al. 1970; Seo et al. 1971; S.W. Kim et al. 1972; N.Y. Park et al. 1973; C.G. Kim et al. 1974; C.G. Kim et al. 1975; J.H. Chang et al. 1978; S.W. Kim and Chang 1979; S.W. Kim et al. 1980a; S.W. Kim and Min 1981; C.S. Kim et al. 1982). Its existence is due largely to the relatively higher discharge of fine sediments from the Keum (Kum, Geum) River, Yeongsan River, and adjacent land influence in excess of coarse-grained sediments (H.I. Choi and Hahn 1975; D.C. Kim 1980; J.H. Choi 1981; S.C. Kim 1982).

Poorly sorted muddy sediments also prevail in the southwestern tip of the peninsula except in the nearshore area and around islands where strong tidal currents winnow fine materials. Tidal currents mold the sandbars that consist of very well to moderately sorted sands in the nearshore area such as near the Odo Island (S.W. Kim et al. 1980a; Klein et al. 1982). In the Jeju and Korea straits, however, the sediments consist mainly of well to poorly sorted sand and gravel (Y.A. Park and Song 1971; S.W. Kim et al. 1980b; S.W. Kim and Min 1981; Suk 1981) (fig. 2.20). This is due to the strong tidal currents and also probably the influence of the Tsushima Current (a branch of the Kuroshio) flowing eastward after hugging the Jeju Island from the south (K. Kim 1980). The effect of the Kuroshio north of Jeju Island has also been demonstrated by the faunal assemblages both in the water column (Y.C. Park 1981) and in the bottom sediments (Choe 1981).

As shown on seismic profiles (fig. 2.16), the sediment layer in this area is thin and often the basement rocks are exposed on the sea floor. The existence of iron-stained quartz sands found in sandy and gravelly sediments (S.W. Kim and Min 1981) indicates that the postglacial accumulation of sediment has been minimal. Rather the sediment has undergone erosion locally.

Mineral and Geochemical Composition

The bulk of sediments in the Yellow Sea consists largely of detrital fractions with minor amounts of insoluble grains of organic and authigenic origin. The calcium carbonate is about 10% and increases southeastward (fig. 2.19c). The amounts of terrigenous silt and clay are dominant (57.8%) in the west, whereas sand components (mainly light minerals) are dominant

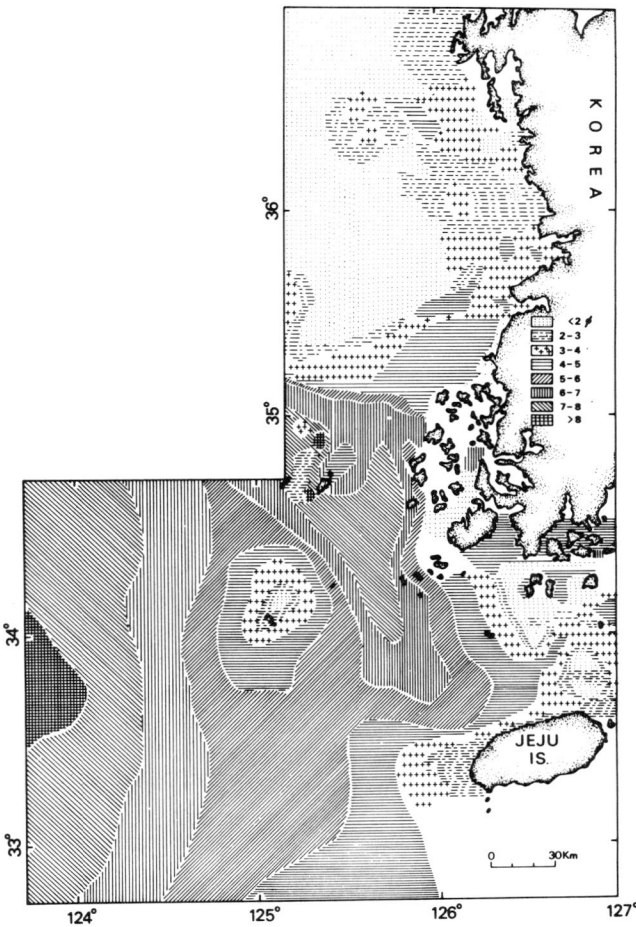

Figure 2.20. Generalized map showing mean grain-size distribution in the southeastern Yellow Sea. Modified after Chough and Kim (1981) by permission of the Journal of Sedimentary Petrology.

in the east. In the east, the light mineral fractions of sand size are subrounded to angular and consist mainly of quartz (43–95%), K-feldspar (3–4%), plagioclase (3–25%), and rock fragments (1–10%) (figs. 2.21a and 2.21b) (N.J. Kim et al. 1970; Seo et al. 1971; S.W. Kim et al. 1972; H.I. Choi and Hahn 1975; C.G. Kim et al. 1975; S.W. Kim et al. 1977; Koo et al. 1980b), pertaining to arkose or lithic arkose (when compacted).

Near the Korean Peninsula heavy mineral assemblage in the fine sand fractions consists of hornblende, olivine, ilmenite, garnet, leucoxene, zircon, epidote, rutile, monazite, and alterite, etc. (total less than about 2.5% of sediment) (Niino and Emery 1961; N.J. Kim et al. 1970; Seo et al. 1971). Also reported are small amounts of orthopyroxene, clinopyroxene, tourmaline, magnetite, apatite, hematite, and sphene (N.J. Kim et al. 1970; Koo et al. 1980b).

Foraminiferal content in the sediments ranges from about 5% in the Yellow Sea to about 40% in the shelf area south of Jeju Island (Niino and Emery 1961; J.J. Kim 1970; C.G. Kim et al. 1975). In the eastern Yellow Sea they are found in abundance in the muddy sediments in the area south of about 35°30′N (J.J. Kim 1970; B.K. Kim et al. 1970; S.K. Chang and Kim 1976). A total of 256 faunal species (87 genera) of both planktonic and benthonic foraminifera are found in the Yellow Sea. Planktonic foraminifera includes both cold-water (and temperate) species such as *Globigerina bulloides*, *G. falconensis*, *G. pachyderma*, *G. quinqueloba* and warm-water (subtropical) species such as *Globigerinoides ruber*, *G. sacculifer*, *Globoquadrina dutertrei*, *Pulleniatina obliquiloculata*, etc. The latter, according to B.K. Kim et al. (1970), are due to incursion of the warm Kuroshio into the southeastern Yellow Sea. Benthonic foraminifera found in the sea in abundance includes *Pararotalia nipponica*, *Ammonia beccarii*, and *Hanzawaia nipponica* which are also found in muddy sediments of less than about 40% sand. Also common are shell fragments, diatoms, ostracodes, spines of echinoids, fish teeth, and other organic remains.

Organic carbon content in the bulk sediment is less than about 0.3% (0.1–0.9%) (fig. 2.19d) and is usually found in the fine-grained sediments. In the nearshore embayments and estuarine environments, it is relatively high, ranging up to 3.0%

Figure 2.21. *Classification of sediment composition according to 3 light minerals (quartz, orthoclase, and plagioclase) in sediments from the Yellow and South seas. a: coastal region; b: off- and near-shore. After Koo et al. (1980b) courtesy of the Korea Ocean Research and Development Institute Bulletin.*

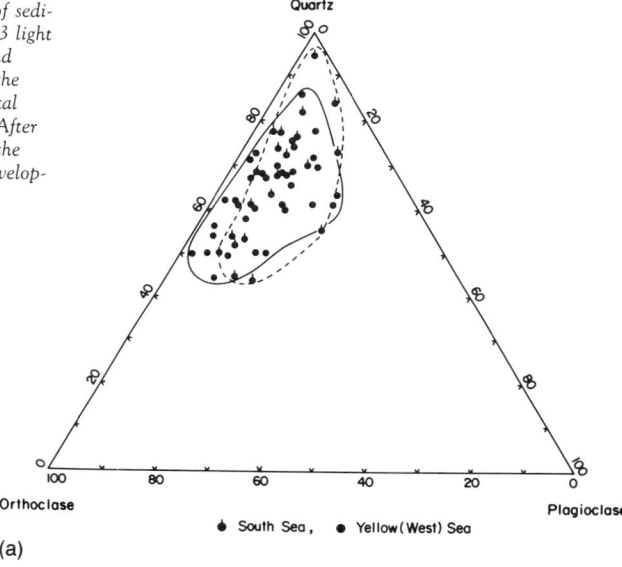

(a)

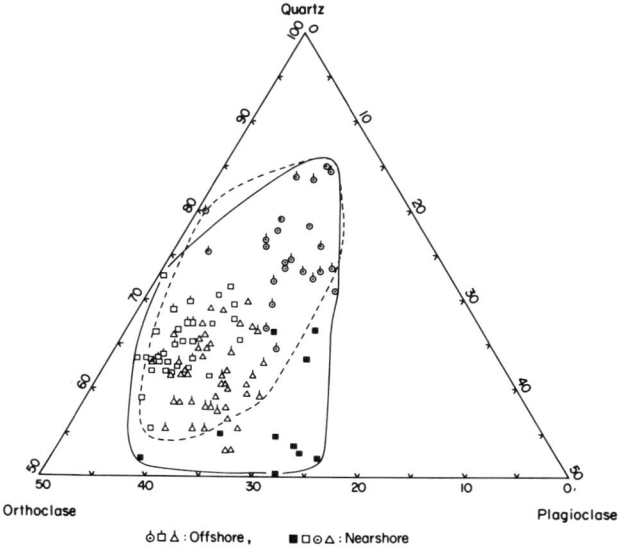

(b)

in the Gyeonggi Bay (S.W. Kim et al. 1979; Koo et al. 1980b) and off the Yeongsan Estuary (J.H. Chang et al. 1978; S.W. Kim and Chang 1979). Nitrogen content in the bulk sediments is approximately between 0.005% and 0.01% (C.G. Kim et al. 1975).

Analysis of inorganic bottom sediments in the eastern Yellow Sea, determined by C.G. Kim et al. (1975) and others, shows that SiO_2 (71–86%) is dominant, followed by Al_2O_3 (6–16%), K_2O (3.2–3.9%), Fe_2O_3 (1.2–3.8%), Na_2O (1.2–2.5%), CaO (0.5–6.7%), and MgO (0.1–2.3%). In the coastal mudflats and nearshore area of the eastern Yellow and South seas, it also includes dominantly SiO_2 (46–88%) and Al_2O_3 (2.8–17.5%) followed by Fe_2O_3 (0.3–6.0%), CaO (0.6–6.7%), Na_2O (1.6–3.8%), K_2O (1.2–3.9%), MgO (0.2–3.0%), P_2O_5 (0.01–0.3%), MnO_2 (0.02–0.2%), TiO_2 (0.05–0.7%) and S (0.005–0.7%) (Koo et al. 1980b). Although of minor amounts, trace elements in these sediments include Cu, Ni, Co, Cr, B, V, Ba, Sr, Pb, Zn, In, Zr, Th, U, and Au.

Figure 2.22 compares the relative amounts of total iron (expressed as Fe_2O_3) plus magnesia (MgO) with those of Na_2O and K_2O found in the Yellow and South sea sediments. Iron and magnesia are found in ferromagnesian minerals such as biotite and hornblende, as well as illite, chlorite, and montmorillonite. K_2O and Na_2O are present in alkali feldspars, muscovite, and illite. The sediments fall largely in the middle part, characteristic of arkosic which are found largely in rapidly filling tectonic settings such as taphrogeosynclines (Blatt et al. 1980). The data on chemical composition thus agree with the mineralogical data mentioned previously.

Dispersal Pattern

The dispersal patterns of the fine-grained sediments in the southeastern Yellow Sea were determined by D.C. Kim (1980), J.H. Choi (1981), Chough and Kim (1981), and Chough (1981a, 1983a) by studying less-than-2 μm fraction clay minerals and some trace elements in the bulk sediments. The clay minerals in the sea are represented by an assemblage of kaolinite-chlorite-illite. Montmorillonite is ubiquitously low in concentration and occurs only in trace amounts.

Figure 2.22. *Chemical composition (Fe_2O_3 plus MgO vs. alkalies [Na_2O and K_2O] in sand and mud from the Yellow, South, and East seas.*

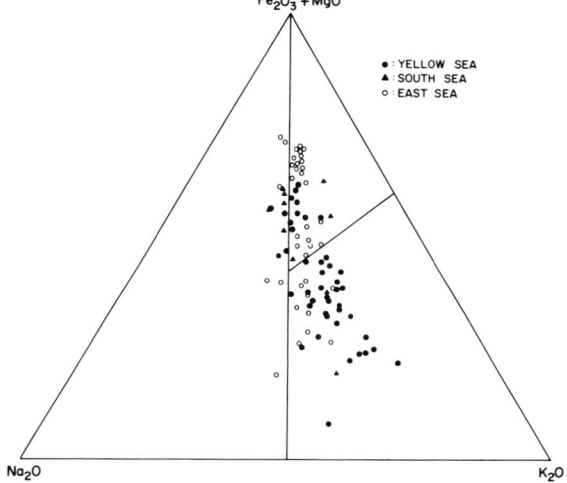

Clay Mineral Distribution

High concentrations of kaolinite of more than 20% occur nearshore, especially near the mouth of Keum River and extend for about 120 km both northwestward and southwestward (fig. 2.23a). The northwest extension follows the shore closely and is deflected southwestward off the Taean Peninsula. In the southwest extension, the zone of high kaolinite is confined to depths less than 20 m. Further south the amount of kaolinite decreases gradually away from the Keum and Yeongsan estuaries.

The chlorite distribution resembles that of kaolinite, but is less distinct (fig. 2.23b). It is generally high in concentration near the shore and at the mouths of both rivers where it amounts to more than 20%.

Illite comprises more than 75% of clays at the expense of kaolinite and chlorite (fig. 2.23c). It increases seaward, away from the influence of the Keum and Yeongsan rivers, concentrated in the area south of 35°N and offshore, west of the Keum Estuary. The bulk of illite in the large portion of the Yellow and East China seas is believed to originate from the Hwangho and Yangtze rivers (Gradusov 1974; Chen 1978).

Distribution of Trace Elements

Of the six trace elements analyzed, four (i.e., Zn, Cu, Ni, and Fe) show the trend of enhanced concentration near the Keum and Yeongsan estuaries and the adjoining nearshore area. The other two (i.e., Pb and Co) are rather scattered (Chough 1983a), different from the distributional pattern of fine-grained sediments found earlier by Chough and Kim (1981).

The distribution of Zn (fig. 2.24a) shows some similarities with those of kaolinite and chlorite: increased values (over 150 ppm) off the west of the Keum and Yeongsan estuaries and the nearshore area within about 50 km from the shore. The Zn values in the bottom sediments at the mouth of the Keum River determined by W.Y. Kim and Park (1978) range from about 60 to 170 ppm. Koo et al. (1980b) also reported the similar values in the suspended particulate matter that range between 15 and 50 $\mu g/l$, with an average of 25.5 $\mu g/l$ (K.W. Lee et al. 1979). The distribution of Zn is associated probably with the influx of material from the Precambrian gneiss and schist in the drainage

Figure 2.23. *Distribution of clay minerals (less-than-2 μm, carbonate free fractions) in the southeastern Yellow Sea. a: kaolinite; b: chlorite; c: illite. Dots: sample stations. Modified after Chough and Kim (1981) by permission of the Journal of Sedimentary Petrology.*

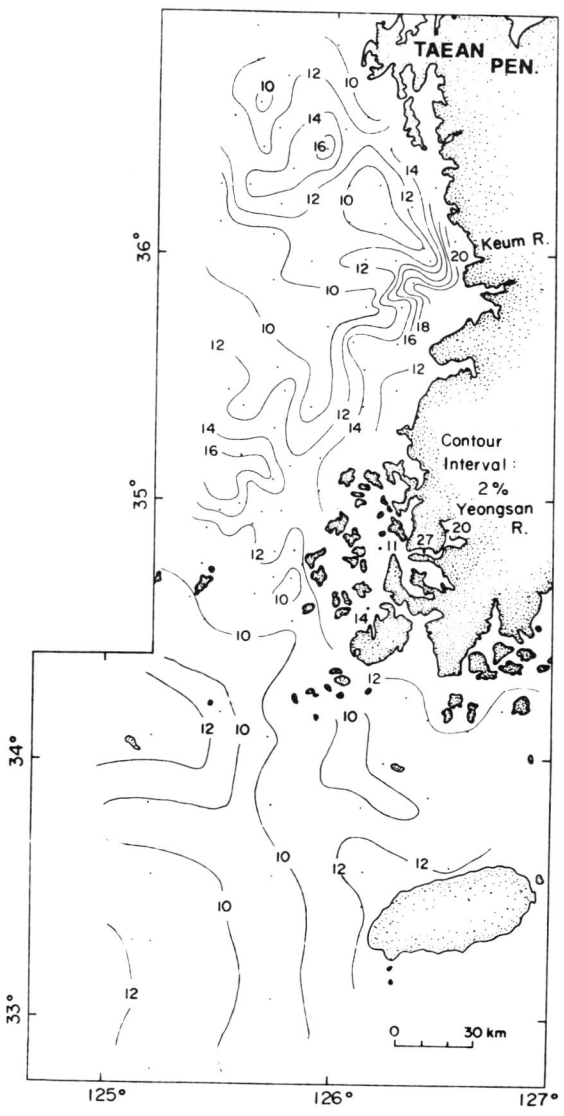

(a)

48   Marine Geology of Korean Seas

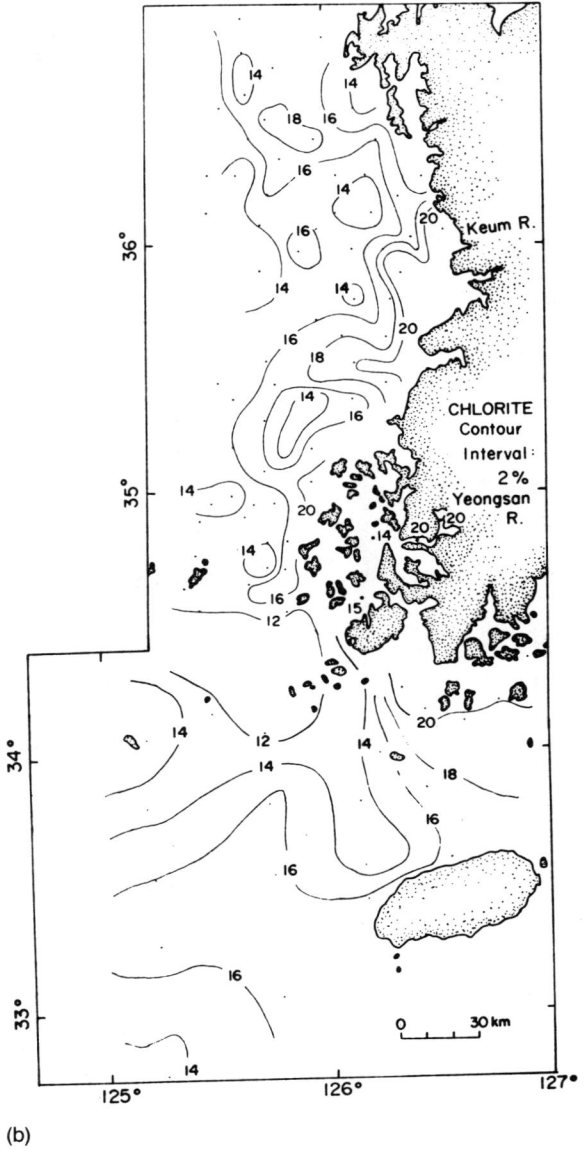

(b)

Figure 2.23b.

Epicontinental Sea (Yellow Sea) 49

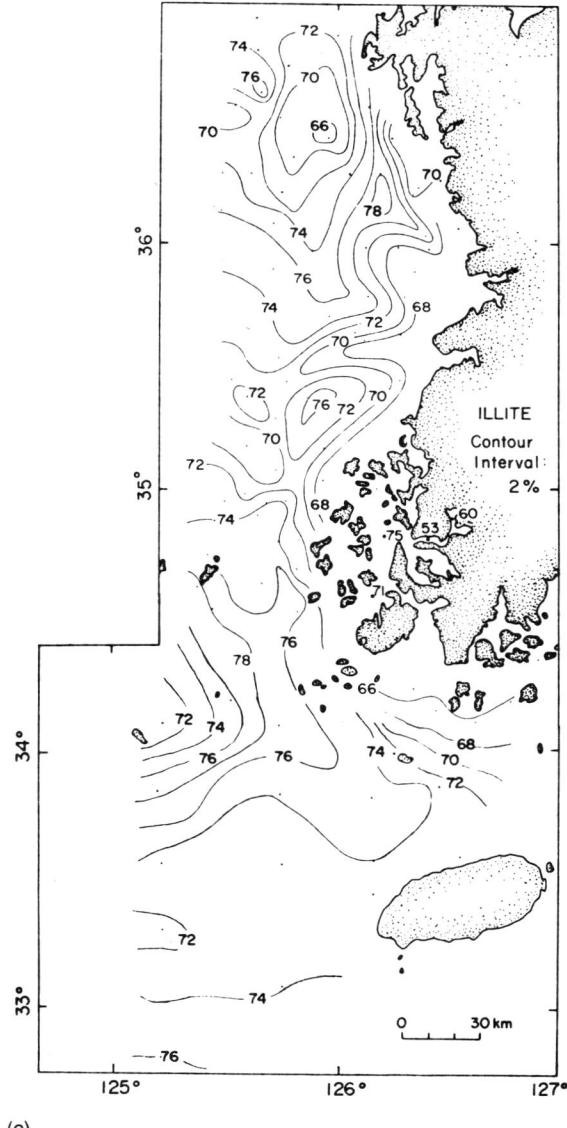

(c)

Figure 2.23c.

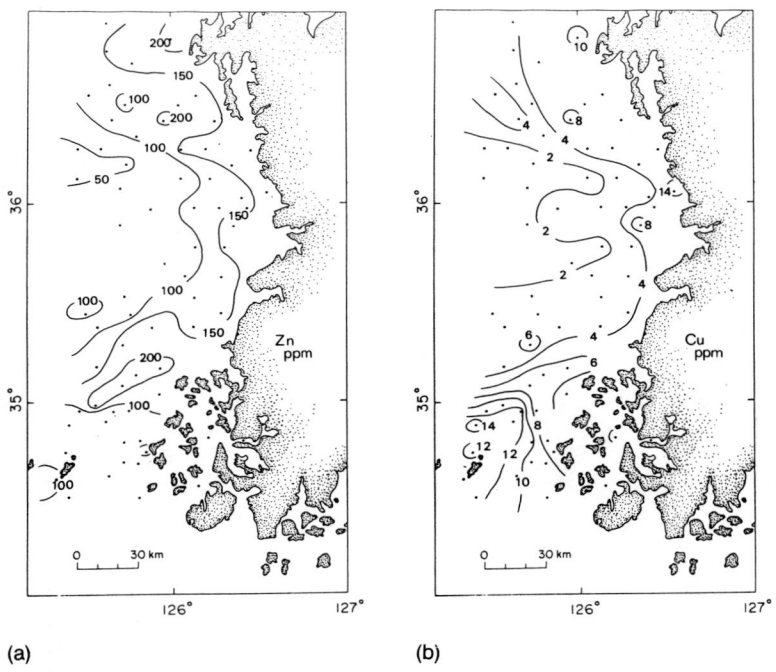

Figure 2.24. Concentration of trace elements in the bulk sediments from the southeastern Yellow Sea. a: zinc (Zn); b: copper (Cu). After Chough (1981a, 1983a) by permission of the CCOP Technical Bulletin.

basins of the adjoining rivers. The Zn concentration in the southeastern Yellow Sea is slightly higher than the world average of about 50–100 ppm in soils (Aubert and Pinta 1977). It is generally high in the muddy sediments (cf. figs. 2.20 and 2.24a), suggesting that the Zn is mainly associated with clay minerals and their colloidal material.

The distribution of Cu concentrations (fig. 2.24b) may reflect sediment yield from the acidic rocks in the drainage areas of the Keum and Yeongsan rivers. The distribution pattern also follows that of Zn, being high near the Keum and Yeongsan estuaries. In the suspended matter at the mouth of Keum River, K.W. Lee et al. (1979) found an average of more than 20 $\mu g/l$ of Cu and ascribed the higher concentration (compared to other estuaries in Korea) to the pollution in the upstream of the river. The general decrease offshore in the west of the Keum Estuary is correlated with the sand content of the sediments and appears to indicate that Cu is mainly associated with clay minerals and other colloidal phases. The nearshore enrichment of Fe concentration (Chough 1983a) may also be explained as the result of Fe-rich clastic sediment influx derived from metamorphic rocks. The Fe concentration in the bottom sediments of the mouth of the Keum River is generally higher than about 25,000 ppm and it is up to 3600 $\mu g/l$ in suspended matters (W.Y. Kim and Park 1978). The iron concentration in the southeastern Yellow Sea is also associated with muddy sediment and may be present as finely divided oxides and hydroxides.

The distribution of Ni is rather low, due probably to the low concentration in acidic rocks and leached podzolic sediments in the temperate region. However, it tends to be concentrated in the nearshore area, with an increasing trend southwestward.

The distributional pattern both of Pb and Co deviates from the above. Pb is concentrated in the upper and lower thirds of the southeastern Yellow Sea whereas the Co concentration is rather patchy. The Co content in acidic or metamorphic rocks is comparatively low, 5–30 ppm (Aubert and Pinta 1977), and the lower content of Co in the southeastern Yellow Sea may be due to its dispersion. The Pb values in the suspended matter in the Keum Estuary and the adjoining southern nearshore range from 5 to more than 20 $\mu g/l$, which is due to the pollution upstream (K.W. Lee et al. 1979).

Dispersal of Fine Sediments

The isotherm patterns compiled by Hahn et al. (1978a, 1978b) seem to agree with the observed distributional pattern of the fine-grained sediments. That is, the horizontal temperature distributions at the fixed-depth intervals observed during the period between 1961 and 1975 reveal that in winter, typically in February (fig. 2.25a), the low temperature (3–6°C) isotherms extend southeastward at all depths, converging shoreward near Taean Peninsula. These isotherms then extend southwestward below about 35°N, which indicates the existence of clockwise nearshore currents in the eastern Yellow Sea. In summer the trend is reversed. The isotherms, especially at 30 m and 50 m depths, extend from the south, suggestive of a counterclockwise current gyre (fig. 2.25b, bottom). A series of *in situ* current measurements in the area slightly north of 35°N made during the period between July 1979 and June 1980 (Nam and Seung 1980) also show the net southwestward movement of about 4cm/s in winter. It is largely reversed in summer. In spring and autumn the temperature structure is transient due to the unstable weather conditions.

Chough and Kim (1981) suggested that the northwestward extending belt of fine-grained sediments rich in kaolinite and chlorite results from the transport of fine fractions out of the Keum River by the counterclockwise nearshore currents during summer, whereas the clockwise circulation during winter is responsible for the southwestern belt of fine-grained sediments (fig. 2.26). The sediment rich in illite in the southwestern tip of the peninsula and the rest of the sea seems to be derived largely from the Hwangho and Yangtze rivers. The influence of the Kuroshio in transporting illite-rich sediments into the southeastern Yellow Sea is considered minimal. The effect of tidal currents on the dispersal of muddy sediments has yet to be evaluated.

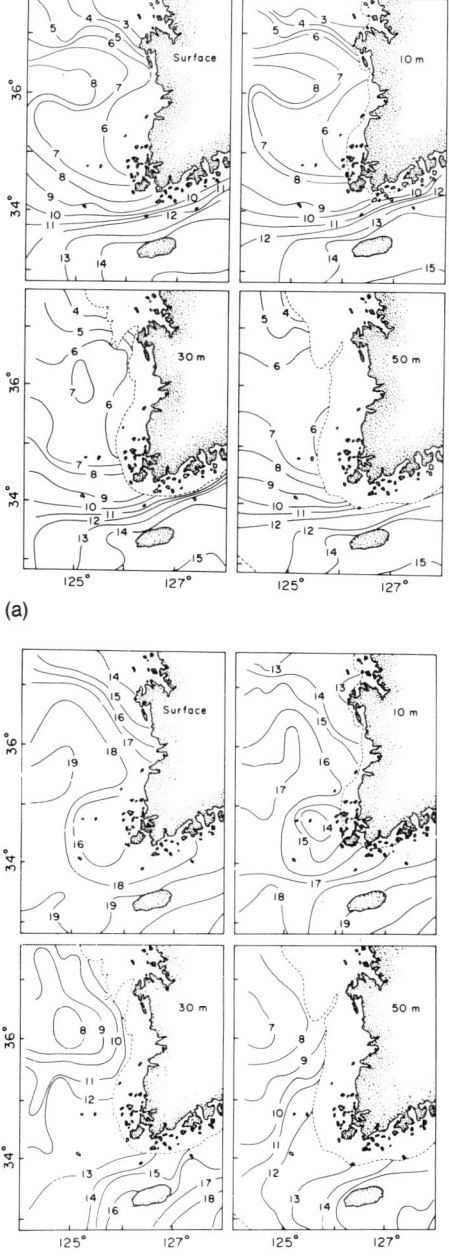

Figure 2.25. Temperature (°C) in the southeastern Yellow Sea. a: February; b: June. After Chough and Kim (1981) by permission of the Journal of Sedimentary Petrology.

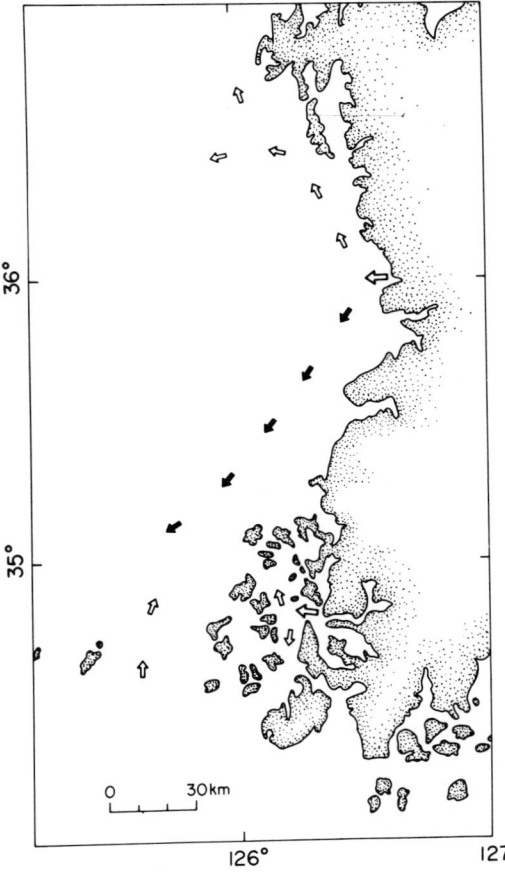

Figure 2.26. Possible transport of the fine-grained sediments in the southeastern Yellow Sea. Modified after Chough and Kim (1981) by permission of the Journal of Sedimentary Petrology.

3 Marginal Sea (East Sea)

Geologic Setting

The East Sea (Sea of Japan) is a typical example of marginal seas or back-arc basins in the western Pacific. The sea floor has been mapped by Terada (1934), Zenkevitch (1959, 1961), U.S. Naval Oceanographic Office (1969), and Mogi (1979). There are three deep basins, the Japan Basin, Yamato Basin and the Ulleung (Ulneung, Ullung, Ulreung; Tsushima) Basin, in the sea (fig. 3.1) which are separated by ridges such as the Korea Plateau, Oki Bank, Yamato Ridge, and the Kita-Yamato Ridge that rise to within about 500 m of sea level. The sea floor in each basin is rather smooth except for a few seamounts and hills.

According to Uyeda and Miyashiro (1974) and Hilde et al. (1976), the East Sea was formed by extension as a result of collision and subduction by the hypothetical Kula-Pacific Ridge from late Cretaceous to Oligocene time. Uyeda (1979) among many others proposed that the sea was rifted by a tensional force, common in Mariana-type subduction, caused by the heat produced from a small-scale convection in the asthenosphere wedge located above the subducting lithosphere (fig. 3.2). Earlier, Bersenev (1971) and Karig (1971) proposed a hypothesis that is similar to the above but suggested a discontinuous diapiric rise of mantle material. Hilde and Wageman (1973) postulated that the opening occurred along two spreading ridges forming the Japan Basin first and followed by the formation of the Yamato and Ulleung basins.

The spreading started either in the late Cretaceous (Bersenev 1971; Murauchi 1971; Hilde and Wageman 1973; Uyeda and Miyashiro 1974) or in Paleogene time (Melankholina and Kovylin 1977). Although the initial rifting may have started in late Mesozoic-Paleogene time followed by nonmarine deposition, the age of the oceanic basement determined by an extrapolation of sediment accumulation rates in the Deep-Sea Drilling Project (DSDP) site 299 in the northeast Yamato Basin indicates late Oligocene to early Miocene (25 to 30 Ma) for opening (Karig and Ingle et al. 1975). This age agrees also with an initiation of subsidence of the Japanese Island Arc associated with the subduction in the Japan Trench (Langseth et al. 1981). The high heat flow in the sea also supports this timing of back-arc opening.

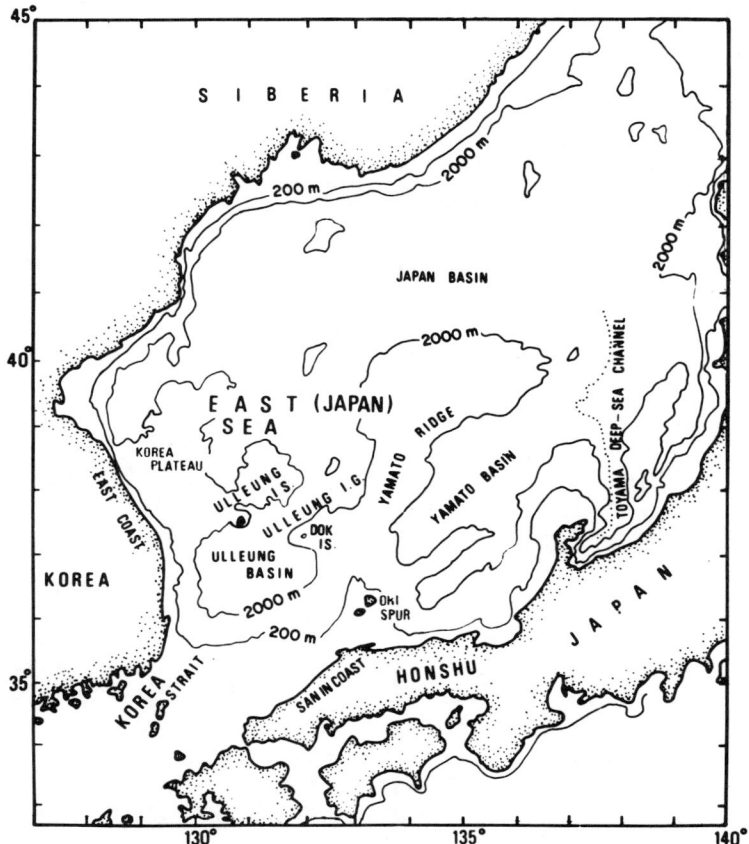

Figure 3.1. Major physiographic features of the East Sea (Sea of Japan). (Note: I.G. stands for interplain gap.)

Marginal Sea (East Sea) 57

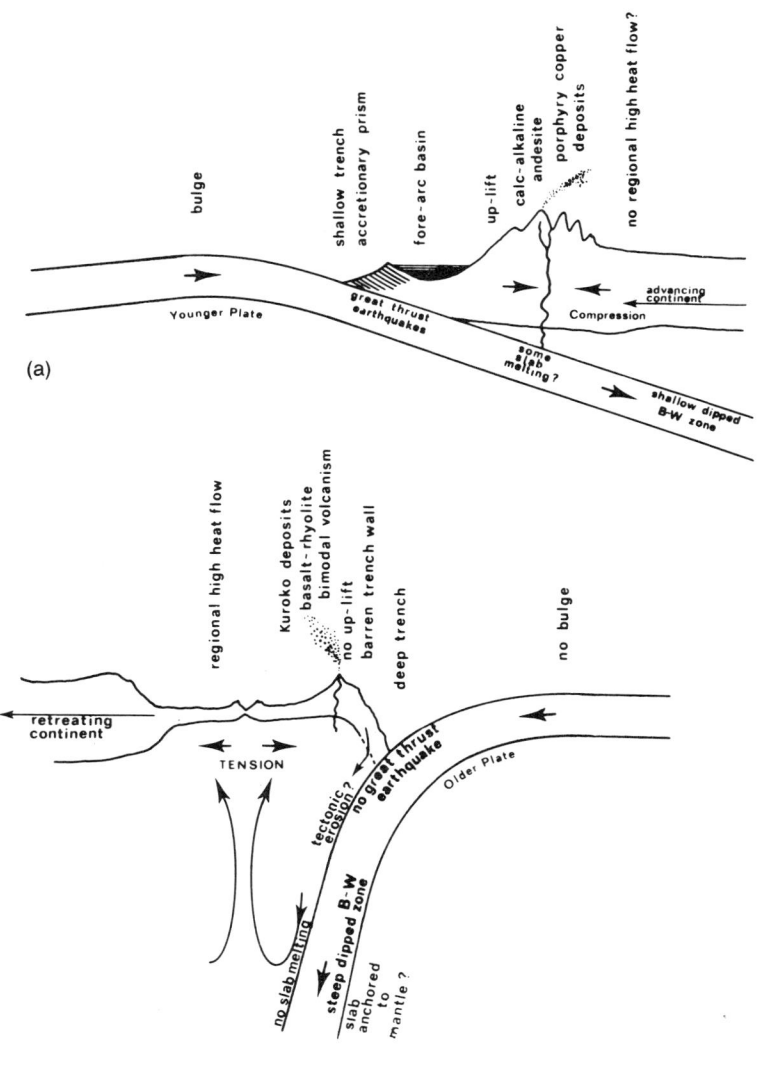

Figure 3.2. Two modes of subduction: Chilean type (a) and Mariana type (b) with back-arc opening (not to scale). After Uyeda (1979) by permission of Oceanus.

Massive deposition of green tuff in Neogene (mainly Miocene) strata along the western coast of Honshu (Minato et al. 1965; Gorai 1968) is viewed as evidence of major rifting in the sea (Ingle 1975). Green tuff was also identified on the Kita-Yamato Ridge (DSDP Hole 302) at the depth of 530 m below the sea floor (Shimazu 1979).

The hypothesis of oceanization of continental crust assumes a possible continuity of geological features found on land under the sea. The basement rocks in the islands and rises of the sea are associated with thick continental crust heterogeneous in age. Gnibidenko (1979) suggested that this represents different stages of geosynclinal basin development since Paleozoic or Mesozoic time. The crust in the sea, in general, consists of a thin veneer of granitic layer and a rather thick basaltic layer underneath. Minato et al. (1965), Gorai (1968), Kaseno (1971), and Minato (1973) argue that the thin granitic layer resulted from erosion of thick Precambrian crust since the Variscan Orogeny followed by subsidence, forming the sea in Neogene time (early Miocene). Melankholina and Kovylin (1977) and Gorai (1982) among others now emphasize the view of lateral migration of the Japanese Islands associated with an injection and effusion of mantle material.

Velocity Structure

Seismic refraction measurements made in the East Sea (Kovylin and Neprochnov 1965; Murauchi 1966; Ludwig et al. 1975) indicate that the deep basins consist of oceanic crust (fig. 3.3). The acoustic basement (layer 2) occurs at a depth of 4 km below sea level in the Ulleung and Yamato basins and at about 6 km below in the Japan Basin (fig. 3.3). It consists either of 3.5 km/s (consolidated sediments or green tuff) or 5.8 km/s layer, the latter being the basement rocks exposed on the adjacent margins (Ludwig et al. 1975). Layer 1, the sediment layer (1.6 to 3.2 km/s), is up to 2 km thick and thickens toward the shelves. In the deep basins it consists of up to 1 km thick undeformed Quaternary turbidites underlain by slightly deformed Pliocene and older sediments. Miocene sediments (or rocks) are block faulted (revealed on the

Marginal Sea (East Sea) 59

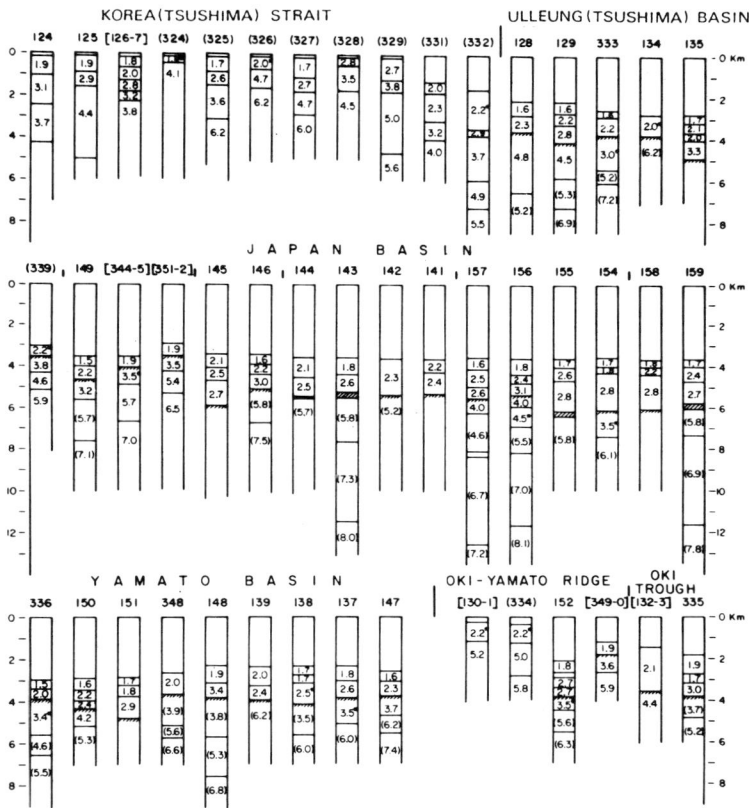

Figure 3.3. Velocity structure sections of various geologic provinces in the East Sea. Velocity is in km/s. Hatchure lines denote acoustic basement. After Ludwig et al. (1975) by permission of the Geological Society of America Bulletin.

60   Marine Geology of Korean Seas

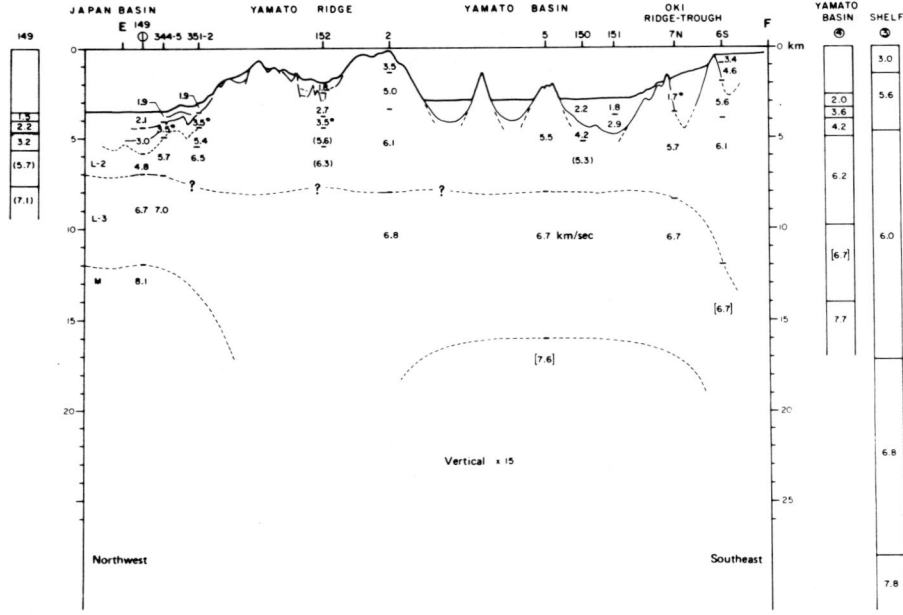

Figure 3.4. Schematic geologic structure (layers 1, 2, 3, and Moho) across the East Sea interpreted from bathymetry, reflection, and refraction profiles. After Ludwig et al. (1975) by permission of the Geological Society of America Bulletin.

margin) and accompanied by volcanism. This will be described later.

The oceanic layer 3 (greater than 6.7 km/s), composed of basalts and metabasalts (Gnibidenko 1979), lies at about 6 to 7 km below the sea level in the Yamato and Ulleung basins and at about 8 km in the Japan Basin (fig. 3.4) (Ludwig et al. 1975). The depth to the upper mantle in the Japan Basin is about 12 km below the sea level. Based on gravity data, Yoshii (1973) estimated the depth to the mantle beneath the Yamato Ridge to be about 23 km.

Magnetic and Gravity Anomalies

Possible spreading centers were located approximately along the long axis of the Japan Basin and of the Yamato and Ulleung basins offset by transform faults (fig. 3.5) (Hilde and Wageman 1973; Isezaki 1975). Based on the prominent magnetic lineations found in the northwestern coast of Japan, Honza (1979a, 1979b) assumed a clockwise radial spreading of the sea with an axis of spreading near the southern coast of Korea. On the other hand, numerous strike-slip faults trending N–S (and NNW–SSE), NE–SE (and NNE–SSE), and NW–SE in the Asian continent (Sikhote-Alin Range and Korean Peninsula) and Japanese Islands led Otsuki and Ehiro (1979) to propose that the sea was formed by a southward drift of Japanese arc during Paleogene time. The drift was bounded by the Tsushima fault on the west and by the Tanakura fault on the east.

The magnetic anomalies compiled by Yasui et al. (1967), Isezaki and Uyeda (1973), and Isezaki (1975) indicate that the amplitude of the anomalies is smaller and the wavelength is about one third of that in the western Pacific. As a result, the lineations of 200 gamma in the anomaly pattern is difficult to recognize and it is not possible to correlate the anomaly pattern with the dated magnetic anomaly time scale in the ocean basins.

The free-air anomaly in the sea is generally positive, 10 to 20 mgal, except for the small minimum near the foot of continental margin or ridge (Stroev 1971; Ludwig et al. 1975; Joshima 1978). This suggests that the sea floor is in isostatic equilibrium and spreading has ceased. Large positive anomalies of up to +50

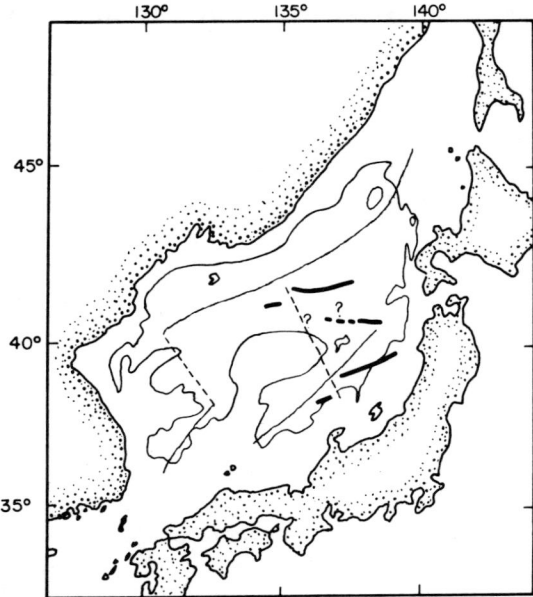

Figure 3.5. *Possible spreading centers (heavy lines) in the East Sea (Sea of Japan) interpreted from magnetic anomalies—Thin lines, possible spreading axis proposed by Hilde and Wageman (1973), broken line for possible transform fault. After Isezaki (1975) by permission of the Journal of Marine Geophysical Research, D. Reidel Publishing Company.*

mgal were observed near such islands as Oki and Koshiki and the Korea Plateau. Negative anomalies (−20 mgal) were observed in the deeper part of the sea. In the Ulleung Basin and its southern margin near Korea Strait, the free-air anomaly is as low as −30 mgal (Joshima 1978).

Heat Flow

Heat flow is large in all the basins of the East Sea, particularly toward the south, reaching 3 HFU in the Yamato and Ulleung basins (fig. 3.6). It is depressed over the intervening ridges between the basins (Yasui and Uyeda 1972; Watanabe et al. 1977). The heat flow falls off gradually eastward to the Japanese islands but drops abruptly along the continental margins of Korea and Siberia. The large flow values are interpreted as a residual effect of an earlier phase of spreading, because no shallow earthquakes occur in relation to the suspected spreading ridges. As mentioned before, the positive free-air anomaly suggests that the sea floor is in isostatic equilibrium and spreading has ceased. The spreading occurred between about 60 and 20 Ma. The existence of alkali-olivine basalts dredged from the flank of the Yamato Ridge (21 Ma) supports this hypothesis (Kobayashi and Isezaki 1976).

According to Schlanger and Combs (1975), the high heat flow in the basin gives rise to kerogen-hydrocarbon transformation in young and shallow strata. However, Kobayashi and Nomura (1972) identified ferromagnetic iron sulfides in cores from the sea, which suggests a stagnant environment during the glacial periods. They suggested that the oxidation of sulfides as well as sulfuration of oxides accounts for the high heat flow in the deep basins of the sea.

Acoustic Stratigraphy

Basement

Parts of the acoustic basement rocks crop out in the Yamato and Kita-Yamato ridges, Korea Plateau, and other seamounts (fig. 3.7). Age determinations of dredged ridge rocks on the various

64 Marine Geology of Korean Seas

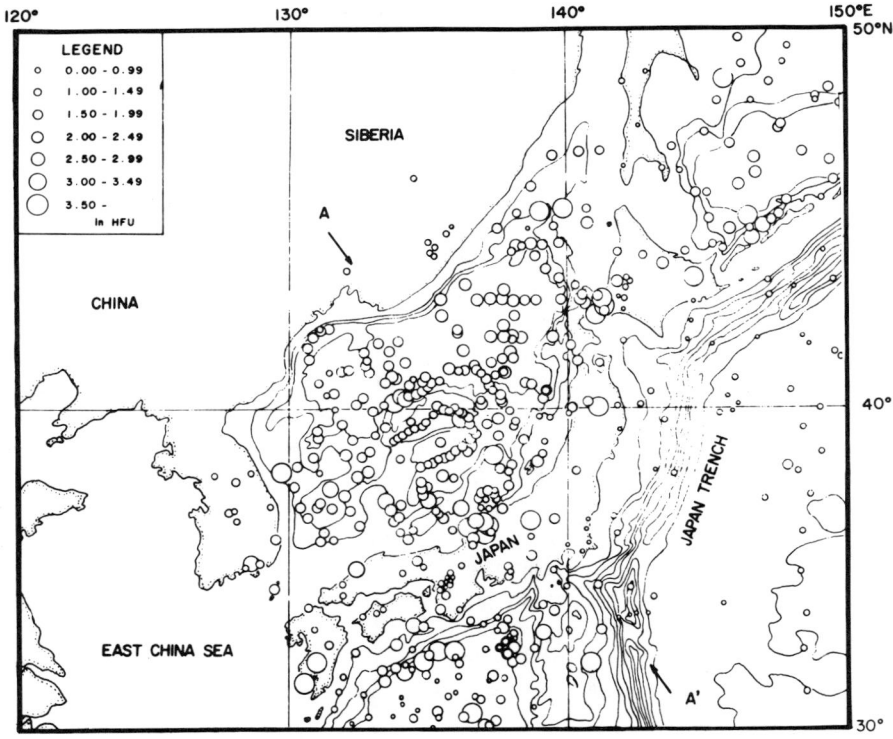

Figure 3.6. Heat flow values in the East Sea and adjacent areas. After Watanabe et al. (1977) by permission of the American Geophysical Union.

Marginal Sea (East Sea) 65

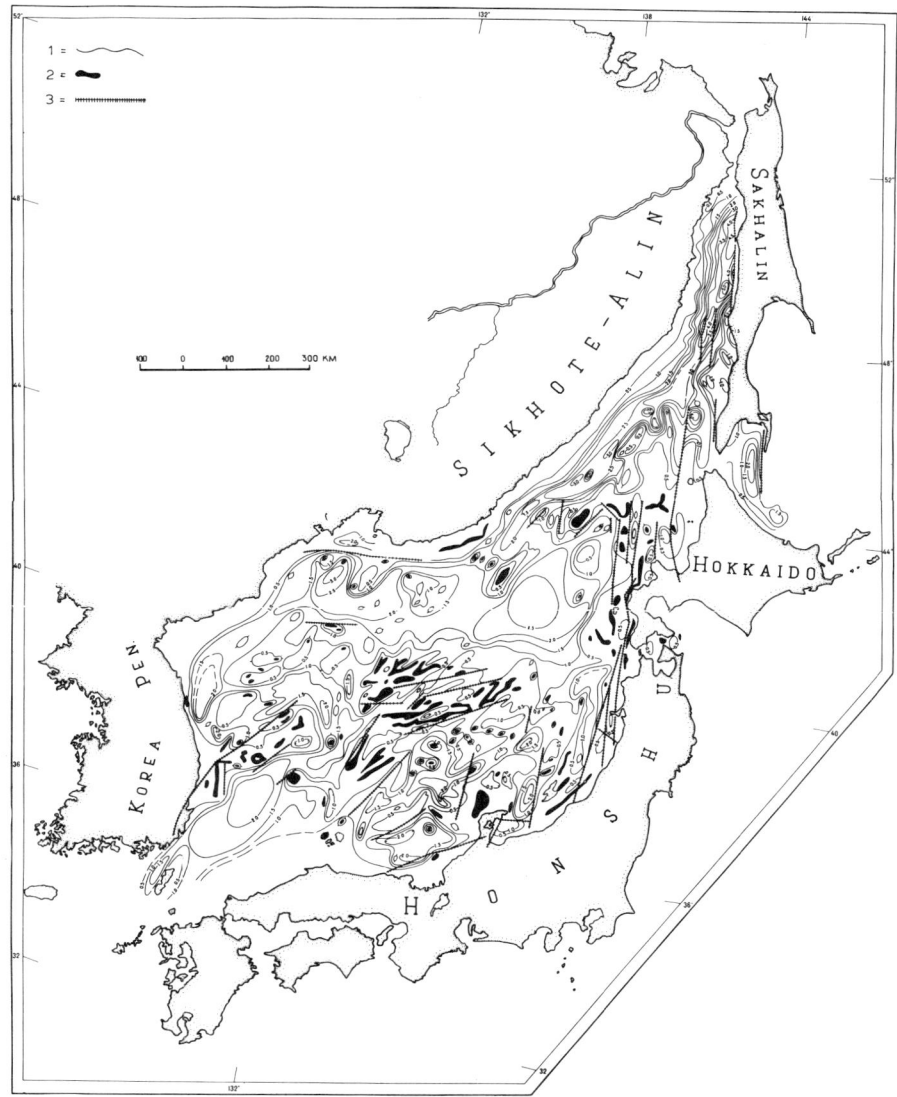

Figure 3.7. Sediment thickness map above the acoustic basement in the East Sea. Contour interval (1) is in 0.5 sec two-way traveltime, assumed average seismic velocity, 2 km/s; dark area (2) for outcrops of acoustic basements (3.5–6.2 km/s); line with cross bars (3) for faults. After Gnibidenko (1979) by permission of Marine Geology, Elsevier Scientific Publishing Co.

banks and seamounts (Hoshino and Homma 1966; Ueno et al. 1971; Lelikov and Bersenev 1975; Gnibidenko 1979) reveal that they are composed both of old (generally older than 60 my) igneous and metamorphic rocks (granite, diorite, rhyolite, andesite, dacite, schist, and gneiss) and relatively younger (less than about 60 my) sedimentary (sandstone and shale) and volcanic rocks (basalt, tuff). Metamorphic rocks of amphibolite facies, various granite-gneiss dredged on the Korea Plateau north of the Ulleung Basin, range in Rb-Sr age from 2729 my to 1983 my (Lelikov and Bersenev 1975). K-Ar ages of the same samples turn out to be younger (180–250 my) which, according to Lelikov (cited in Gnibidenko 1979), is due to thermal events during granite intrusion in Cretaceous time. K-Ar ages of granite and granodiorite dredged from the Yamato Ridge give 197 my, whereas 20 my is the average age for basalts and andesites in the region. Yuasa et al. (1978) reported the occurrence of biotite granite and gneiss, tuff and tuff breccia, and volcaniclastic sandstone from the Oki Bank slope that contain characteristic reddish-brown olivine and volcanic glass and others. The age of andesitic and basaltic rocks dredged from the seamounts near northwestern Honshu ranges in age from 7.7 to 4.2 my (late Miocene to late Pliocene). Quaternary basalts are also reported in the Ulleung and Dok islands of Korea.

Sedimentary Sequence

The total sedimentary sequence (assumed velocity of 2 km/s) in the East Sea (Sea of Japan) compiled by Gnibidenko (1979) is shown in figure 3.7. The sequence is generally thick in the deep basins ranging up to 2.5 sec and thin in the intervening ridges, rises, and toward margins. It is also thick in some local basins along the margin.

The sedimentary sequence consists generally of upper opaque and lower transparent layers (Hilde and Wageman 1973; Ludwig et al. 1975; Gnibidenko 1979). The upper layer is characterized by well-defined reflectors (fig. 3.8) traced over tens of kilometers. Generally it is horizontally stratified except near the ridges and troughs. This layer consists mostly of interbedded turbidites. The lower, transparent layer is composed dominantly of pelagic sediments, mainly diatomaceous oozes (Karig and Ingle et al. 1975). These are most likely correlative with the middle to late Miocene diatomaceous deposits in a number of Neogene se-

Marginal Sea (East Sea) 67

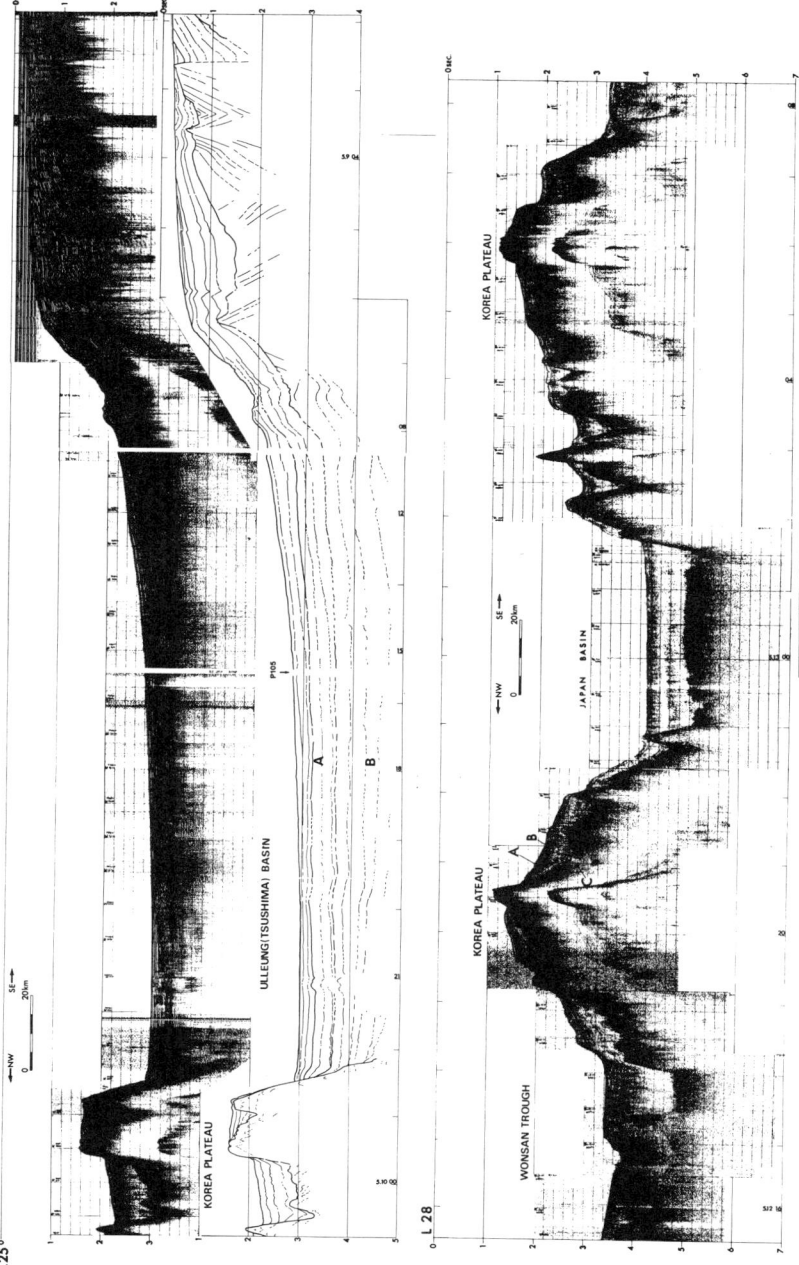

Figure 3.8. Air-gun profiles (lines 25 and 28, for position see figure 5.4). Vertical scale in two-way traveltime in seconds. The upper nontransparent unit (A) underlain by transparent middle unit (B) occurs both in the Ulleung and Japan Basins. In the Japan Basin, the nontransparent unit is also interlayered in the transparent unit (P105, location of piston core). After Tamaki et al. (1978) by permission of the Geological Survey of Japan.

quences along the coast of Honshu and eastern Korea near Pohang (Ingle 1975). The lower layer contains occasionally some opaque layers also (fig. 3.8) (Tamaki et al. 1978). According to the DSDP results (Karig and Ingle et al. 1975) and Honza et al. (1977), the base of the upper opaque layer corresponds to upper Pliocene in age.

The thickness of the opaque layer ranges from 100 m to 1000 m in the Japan and Ulleung basins (Gnibidenko 1979). In the center of the Ulleung Basin south of the Ulleung Island it is about 600 m (Tamaki et al. 1978). Here, the top of the opaque layer is characterized by a smooth surface in the central part, but becomes irregular and tilted both toward the northwestern and southeastern margins. The tilting is gentle toward the southeast but rather abrupt toward the northwest. Piston cores taken recently by Honza et al. (1978) reveal that the uppermost column of this opaque layer consists predominantly of thinly laminated turbidite sequences with alternating minor amounts of hemipelagic sediments (Chough 1982).

Surface Sediments

General Statement

The distribution of sediments on the sea floor was compiled by Skornyakova (1961), Kaseno (1972), and Repechka (1973). Sediments may be classified into gravel, sand, silt, silty clay, and clay. Pelagic components are mainly composed of diatoms (Hasegawa 1970; Koizumi 1970, 1978) and subordinate silicoflagellates (Shitanka et al. 1970). Foraminiferas (and thus $CaCO_3$) in the deep basins are rather rare (Asano 1957; Kozak 1974; Ichikura and Ujiié 1976), also due to the shallow carbonate compensation depth (CCD) of about 2000 m caused by deep circulation of highly oxygenated water (Niino et al. 1969). Terrigenous components of the hemipelagic sediments in the deep basins, originated from the Asian continent and Japanese islands, are also highly oxidized and brownish in color. The influence of the Tsushima Current has also been suggested in the transport of fine-grained materials from the southwest (Aoki et al. 1974).

Distribution (fig. 3.9)
Sandy and gravelly sediments occur along the shallow portion of the sea, some of which were transported by ice in the northern part of the sea (Skornyakova 1961; Niino and Emery 1966). Sandy gravels on the Yamato Rise are composed of rock fragments of volcanic origin including diabase, gabbro-diabase, porphyrite, andesite, and andesite-basalt. Gravels of basic and alkali extrusive rock and some amounts of granite and quartz porphyry also occur (Ueno et al. 1971).

Sandy sediments are confined mainly to a depth of 70 m on the shelf, but also occur extensively along the Korea Strait. The sediments are usually noncalcareous ($CaCO_3$ content is about 0.15–3.04%). Calcareous shell sands are found in the Korea Strait and along the coast of Honshu where $CaCO_3$ content is more than 15%. Calcareous foraminiferal sands occur in abundance (up to 30%) on the Yamato Bank and Korea Plateau dominated by *Globigerina pachyderma*, followed by *G. bulloides*, *G. quinqueloba*, *Globigerinita glutinata*, *Globigerinoides rubescens*, *Globigerinoides ruber*, and others (Kozak 1974).

Coarse silt-size sediments are usually green and greyish green in color and contain abundant shell fragments. They occur mainly on the shallow portions of the sea: shelf, part of the slope, and also near the islands along the coast of Japan. Carbonate content in silty sediment is 0.3–2.1%. Fine silt and clay (0.05–0.01 mm) are dominant in the deep basins of the sea including slope, as well as in some coastal embayments. They are low generally in carbonate content (0.13–1.56%). In the northern part of the sea, the silty clay contains abundant plant fragment and siliceous algae.

In contrast to the Yellow and South seas, montmorillonite occurs in abundance (up to 25%) in the East Sea (Niino et al. 1969; Aoki and Oinuma 1973; Aoki et al. 1974). This is attributed to the influx of weathering products of volcanic, igneous and pyroclastic rocks and soil on the adjoining margin. Kaolinite is less abundant (up to 15%) than illite (up to 50%). Chlorite content is about 30%. It is believed that large amounts of illite originate from the Korea Strait (Han 1979) and partly from the eolian transport by the jet stream. The kaolinite content is about average for the north Pacific region.

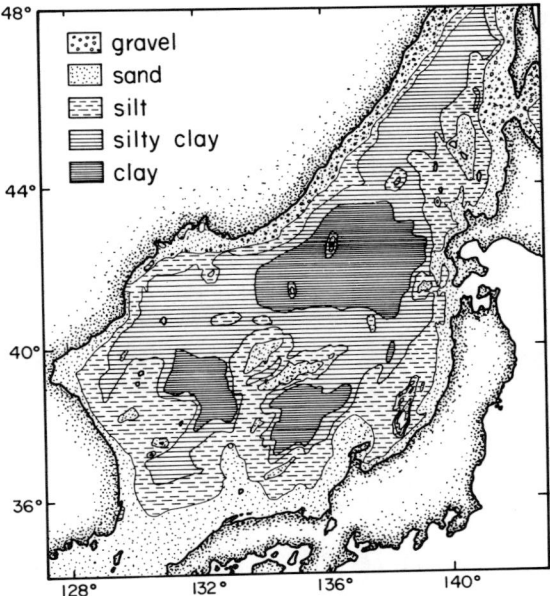

Figure 3.9. Grain size distribution of surface sediments in the East Sea. Modified after Skornyakova (1961).

Geochemical Composition

Coarse-grained and calcareous sediments on the shelves and banks contain minor amounts of organic carbon (Niino and Emery 1966), whereas fine-grained, slightly calcareous sediments on the slope contain relatively large amounts of up to 2% (fig. 3.10) (Solov'ev, cited in Strakhov 1962; Niino et al. 1969). In the basin floor, yellowish brown muds are less calcareous (less than 1% $CaCO_3$) and contain less than 1% of organic carbon.

Chemical composition of sediments, excluding biogenic amorphous silica and carbonate, determined by Sakanoue et al. (1970) and Repechka (1973), shows that silica content is generally high (up to 70%) in sandy sediments, whereas it is less than 50% in pelitic oozes. CaO (2.4–1.2%) and total alkalies are also low in the pelitic sediments, whereas those of Al_2O_3 (11.7–17.6%), MgO (1.2–2.6%), Fe_2O_3 (2.7–3.7%), P_2O_5 (0.08–0.34%), and MnO (0.04–0.73%) are generally abundant in fine-grained sediments. In volcanogenic sediments, the amounts of silica, magnesia, total iron, MnO, CaO, Al_2O_3, and TiO_2 are generally high compared with those in sandy sediments. The relative amounts of total iron plus magnesia compared with those of Na_2O and K_2O in the East Sea sediments are shown in figure 2.22. East Sea sediments fall largely in the high ferromagnesia, characteristic of eugeosynclinal or tectonically active setting.

Late Quaternary Stratigraphy

Late Quaternary stratigraphy in the East Sea cores may be established based on the lithology, biostratigraphy, tephrochronology, and the chemostratigraphic methods using oxygen and carbon isotopes.

Lithology

Ichikura and Ujiié (1976) examined in detail 26 piston cores from the sea taken aboard R/V Vema and R/V Robert Conrad. In the three deep basins, homogeneous and laminated clays are dominant, whereas diatomaceous clay or ooze prevails on the ridges. The topmost sediment column (0–2 m) is usually homogeneous brown clay, whereas the lower portion is commonly laminated. On the Yamato Rise, however, it is replaced

72   Marine Geology of Korean Seas

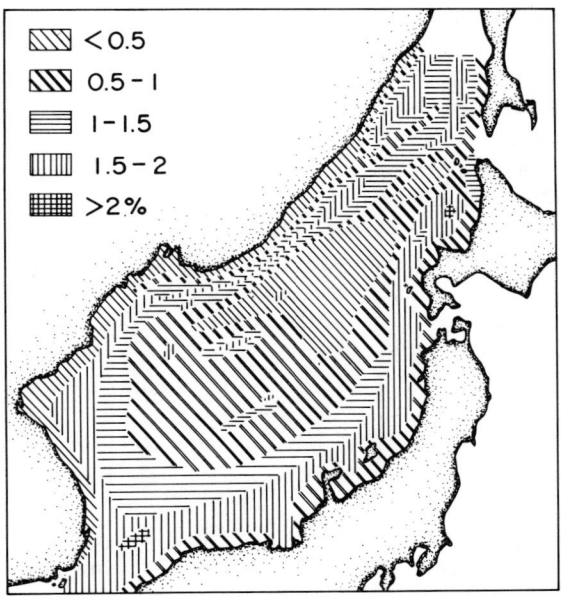

Figure 3.10. *Distribution of organic carbon (%) in sediments of the East Sea. After Solov'ev (cited in Strakhov 1962), Niino et al. (1969) and G.H. Lee (1983).*

by sandy clay composed of quartz, volcanic glass, glauconite, mica, and some mafic minerals, and is usually barren of biogenic components. Homogeneous brown clay is rich in diatoms with subordinate amounts of radiolarians, volcanic glass, and minute quartz grains. Foraminiferal sands occur in the laminated units (turbidites) suggesting that they were transported from the shallower part of the sea by turbidity currents.

Other sediment types found in the sea include glauconitic sands exclusively on the Yamato Rise, sandy clay, and sands. Found exclusively in the deep basins, the latter contain angular to subangular quartz, igneous rock, plant fragments, etc. Microfossils such as foraminifera, sponge spicules, and ostracods also occur in the sands. Sands containing abundant shell fragments are widely distributed in the southern margin of the Ulleung Basin.

Volcanic ash layers are abundant in the sea and contain light olive gray (5Y 6/1) volcanic glasses of quartz, pyrite, and other mafic minerals. Manganese micronodules indicative of oxidizing condition were found in surface sediments in the sea deeper than about 2100 m. Tiny framboidal pyrite (3–25 μm in size) aggregates of octahedral form also were found in the foraminiferal test, radiolarian test, diatom frustule, or in other tubular organisms in the sediments of pre-Holocene age.

Holocene-Pleistocene Boundary

In three deep basins, C-14 dates on the lithologic boundary from homogeneous hemipelagic clay to laminated turbidite occur at about 9000 y B.P. Although many turbidite cores do not contain foraminifera, Ujiié and Ichikura (1973), Ichikura and Ujiié (1976), and Arai et al. (1981) found that the coiling ratio of *Globigerina pachyderma* changes from left to right across this lithologic boundary. This is also accompanied by an extinction of *Globigerina umblicata* which Ujiié and Ichikura (1973) ascribed to a change from a reduced environment to an oxidized one across the boundary. This change was substantiated by an abrupt increase of benthonic foraminifera resulting from an increase of oxygenated bottom water in Holocene time. Manganese micronodules are associated with this environment, whereas framboidal pyrites occur below the boundary when the sea was isolated from the open ocean and thus was a reduced environment.

The depth to the Holocene-Pleistocene boundary not only differs regionally, but also according to the method applied to identify it. It ranges from less than 1 m below the sea floor in the Japan Basin (Koizumi 1970) to about 2.8 m in the Yamato Basin. In the Ulleung Basin, it occurs at about 2 m (Chough 1982). Accumulation rate of hemipelagic sediments above the boundary is on the average 12 cm/10^3 y, which is substantially higher than the rate (1.5 cm/10^3 y) determined by the Io/Th method suggested earlier by Miyake et al. (1968). The data on diatoms and foraminifera (Koizumi 1978) agree with the former.

Detailed works on volcanic ash in core sediments (Mizuno et al. 1972; Arai et al. 1981) indicate that there are several ash layers that are widespread and can be correlated with each other. They include Akahoya ash from Kikai Caldera, Kyushu (ca 6300 y B.P.), Aira-Tn ash from Aira Caldera (ca 21,000-22,000 y B.P.), and Aso-4 ash from Aso Caldera (ca 50,000 y B.P.). In the central and southeastern part of the sea, the marker layers include Ulleung ash (Chough 1982) from the Ulleung Island (ca 9300 y B.P.) and Yamato ash (ca 25,000-35,000 y B.P.).

Paleoceanography

Vertical changes in lithologic, foraminiferal content, and coiling ratio of *Globigerina pachyderma*, and oxygen isotope ratio occur at about 6500-9500, 13,000, and 23,000 y B.P., respectively. Paleoceanographic conditions determined from the values of oxygen isotope ratios in benthonic and planktonic foraminiferal test (Arai et al. 1981) suggested that the water mass in the sea was relatively constant (salinity of 33-34‰) with relatively low temperature (8-10°C) during the period between 60,000 y and 23,000 y B.P., influenced only by minor amounts of seawater from the Pacific. A decrease in salinity accompanied by a decrease in δO^{18} values in foraminiferal test occurred during the last glacial period between 23,000 y and 13,000 y B.P. The increased δO^{18} values of planktonic foraminifera tests at about 13,000 y B.P. signal the inflow of Pacific waters this time. At about 8000 y B.P. warm water (7-8°C) species of foraminifera appeared accompanied by an appearance of dextral-coiling *Globigerina pachyderma*. The latter was also demonstrated by Kato

(1978) in four cores from the Ulleung Basin in which the dextral form replaced the sinistral form at about 10,000 y B.P., corresponding to the climatic change from glacial to interglacial. The number of G. *pachyderma* increased also.

Various lines of evidence presented in the foregoing sections suggest that the East Sea was isolated from the open ocean during the late Pleistocene glacial period when the sea level stood lower. The water in the sea was most likely stagnant and covered with ice during this time lacking vertical circulation of the water masses. The reduced environment of bottom water probably permitted pyrite and sulfide minerals to form on the bottom. Turbidity currents and associated mass flows were also active during this time.

Postglacial sea level rise resulted in a subsequent inflow of a warm water mass (Tsushima Current, a branch of Kuroshio) through the Korea and Tsushima straits, which flows mostly northeastward along the Japan side of the sea. Winter cooling in the north end of the sea caused the formation of the bottom water mass resulting in the high concentration of soluble oxygen in the deep basins (Niino et al. 1969). Due to respiration and oxidation, this bottom water is progressively enriched in CO_2. The CCD, then, is very shallow and the hemipelagic sediments are deficient in calcareous organic remains (Ichikura and Ujiié 1976).

4 Eastern Shelf

Geologic Setting

The eastern shelf of the Korean Peninsula and part of the slope south of 38°N has been surveyed in detail (Schlüter and Chun 1974) using air-gun reflection and refraction methods. Figure 4.1 shows the shelf sea floor (up to 30 km wide). This sea floor is rather smooth and progressively deepens toward the slope except where broad mound or bank, tentatively named here the Hupo Bank (after the port west of it), rises more than 100 m above the surrounding sea floor. The Hupo Bank is due to the large accumulation of sediments dammed seaward by the basement high bound by faults (cf. figs. 4.2 and 4.3).

The shoreline is largely straight and lacks embayments except Yeongil Bay near Pohang where sediments are composed of locally derived, poorly sorted gravel, sand and silty clay derived mainly from the Hyeongsan River (B.K. Park and Song 1972; C.S. Kim and Kang 1973). The coastline is characterized by a number of terraces (ca 6) ranging in height from about 3-7 m to 90-130 m above sea level (S.W. Kim 1973). C-14 ages on the peat and charcoal fragments from two terraces, respectively, suggest an uplift rate of 1.1-1.4 mm/y. At this rate, the highest terrace (90-130 m above sea level) on the east coast could have formed since the last interglacial period. The difference in elevation of these terraces may be interpreted as due to a differential rate of uplift or faulting during epeirogenic deformation.

Offshore, north of 37°N, the sea bed is characterized by a complex series of shallow ridges and rises, Korea Plateau, that extends for about 200 km eastward (figs. 3.1 and 4.4). The plateau consists of a north branch near 39°N and a south branch near 38°N. The plateau appears to be an extension of granite on land intruded in Jurassic time within the Precambrian gneiss. The plateau is overlain by a thin sediment sequence in the troughs between ridges ranging up to 1.1 seconds in thickness. The sequence is either transparent or slightly stratified, suggesting ponded turbidites in the troughs.

Acoustic Stratigraphy

Figure 4.3 shows the results of an earlier seismic reflection survey, interpreted recently in more detail by C.S. Kim (1981)

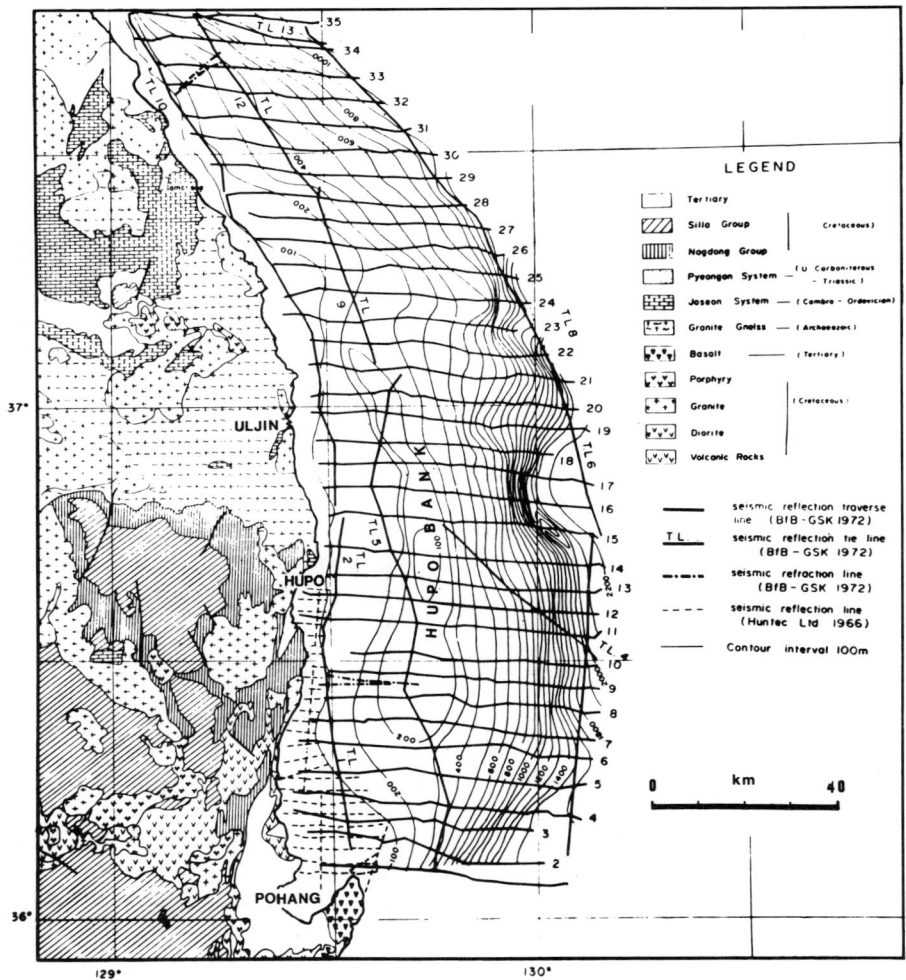

Figure 4.1. Bathymetry and tracklines of seismic survey in the continental shelf and slope of the eastern margin, Korean Peninsula. Land geology is also shown. After Schlüter and Chun (1974) by permission of the CCOP Technical Bulletin.

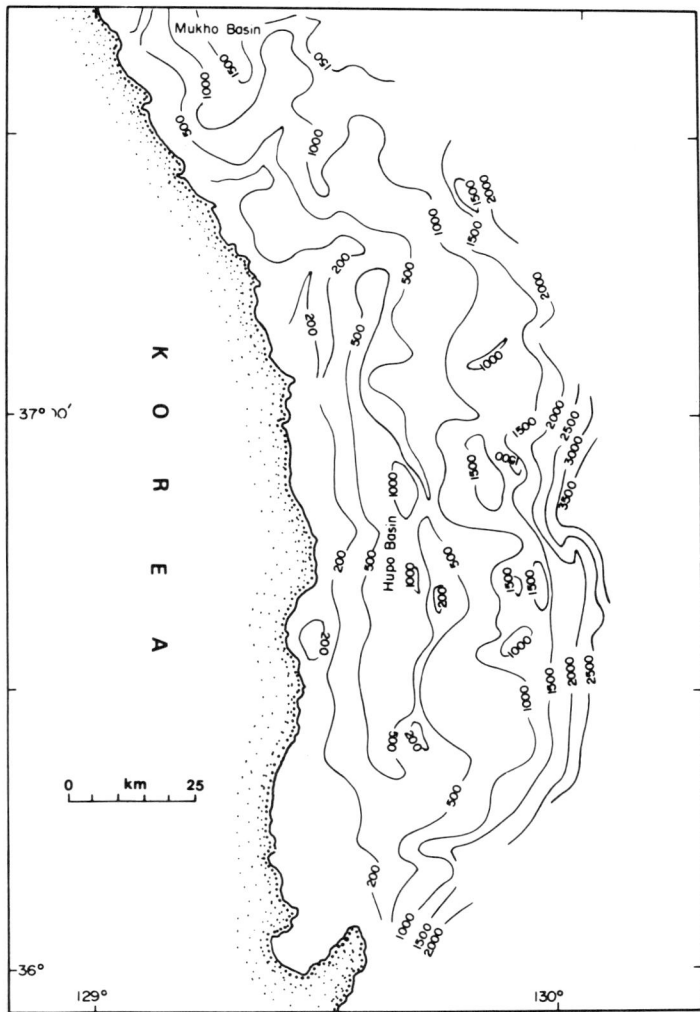

Figure 4.2. Map showing the depth to the acoustic basement from sea level in the eastern margin of the Korean Peninsula. Contours in meters. After K.P. Park et al. (1981) courtesy of the Korea Institute of Energy and Resources Report.

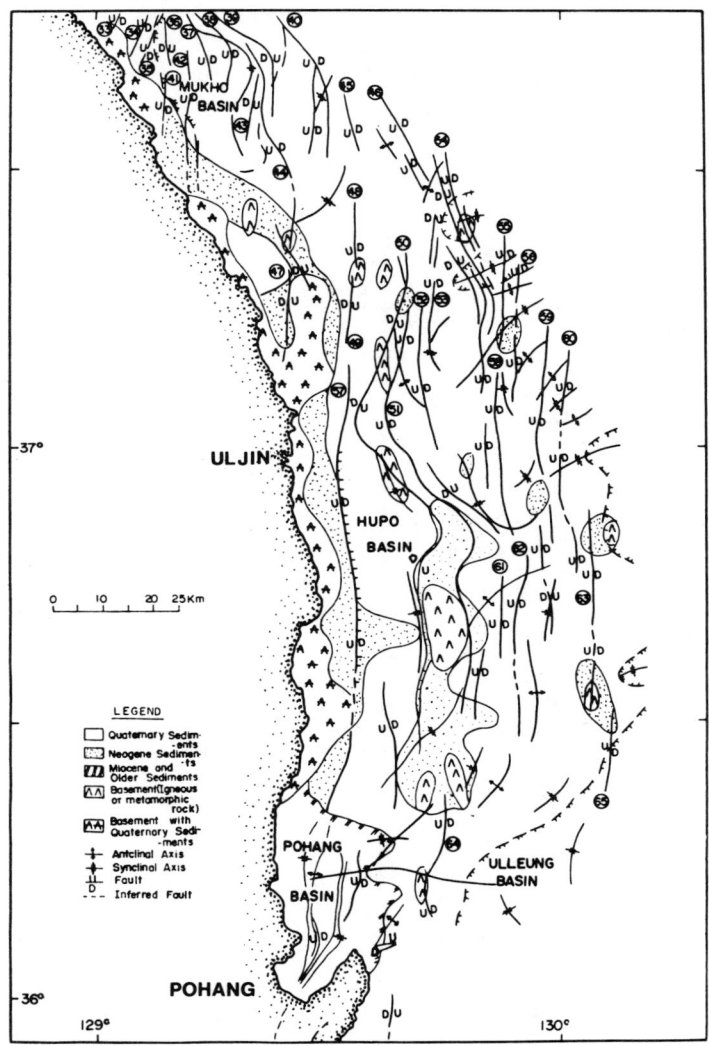

Figure 4.3. Geological map of the eastern margin of the Korean Peninsula, interpreted from the seismic data of figure 4.1. Circled numbers represent the identification of faults (not discussed in the text). After K.P. Park et al. (1981) courtesy of the Korea Institute of Energy and Resources Report.

Eastern Shelf 81

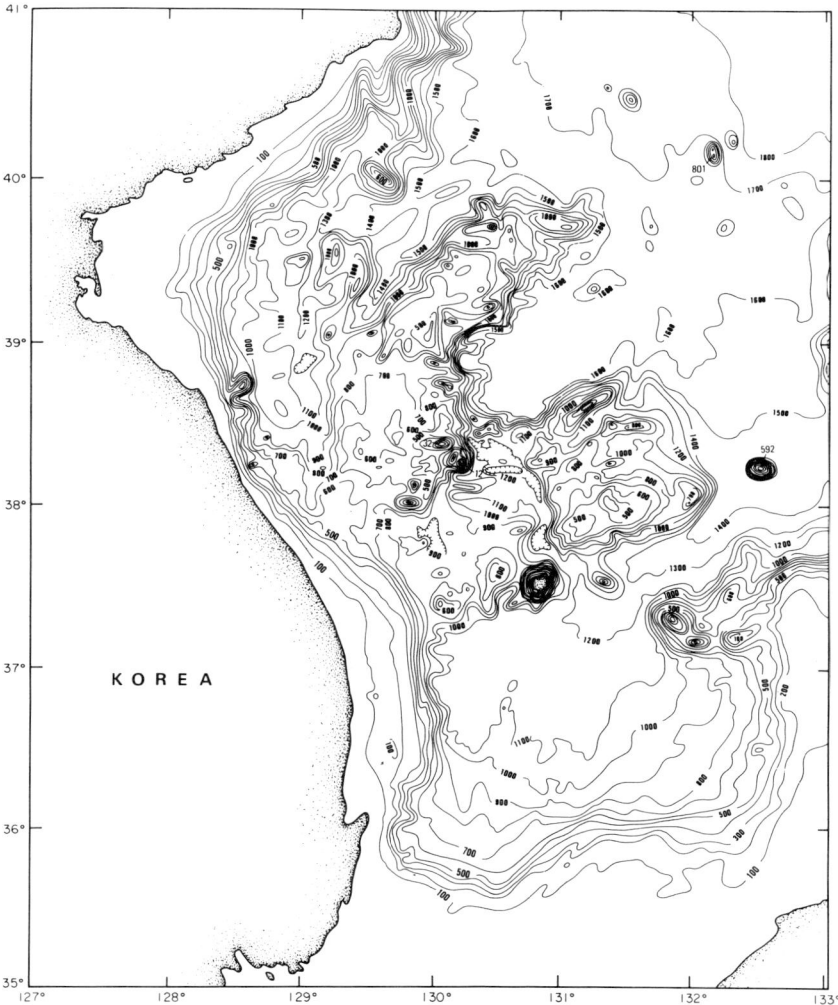

Figure 4.4. *Detailed bathymetric chart of the eastern margin of the Korean Peninsula. Contours in fathoms. This figure represents a part of a map of the East Sea (Sea of Japan) prepared by the U.S. Naval Oceanographic Office, Pacific Support Group (1969). Reprinted by permission of the Department of the Navy.*

and K.P. Park et al. (1981). The eastern shelf is floored by thin Quaternary sediments, underlain by a probable Neogene sequence. Acoustic basement consists of Precambrian gneiss, Cretaceous granite, and sedimentary rocks exposed on land (fig. 4.1). Seismic refraction surveys (Schlüter and Chun 1974) reveal that the interval velocity through the upper Quaternary and Neogene sediments is less than about 2 km/s, whereas through the acoustic basement it is either 4.4 km/s or 5.6 km/s. In the north of Uljin, the acoustic basement consists of nonstratified, 5.6 km/sec (by sonobuoy measurement) rock, which is most likely an extension of Precambrian gneiss. In the south of Uljin, however, the basement is composd of stratified, 4.4 km/s rock and appears to be an extension of Cretaceous sedimentary rocks on land. Neogene sediments are rather thin on the shelf, deposited in two relatively thick basins, the Mukho Basin (about 1400 m thick) and the Hupo Basin (about 850 m thick). Both basins are bounded by numerous block faults trapping postdeformational sediments (fig. 4.5) since the late Miocene. Off Pohang, Neogene sequences probably do not exceed 700 m in thickness (Huntec Ltd. 1968). The Miocene and older rocks in the Pohang Basin are correlated with those along the coast (B.K. Kim 1965).

Detailed seismic surveys and exploratory boreholes in the southeastern shelf off San-in, northeast of the Tsushima Island, reveal that the present shelf is underlain by a thick (more than 4 km in thickness) sequence of Tertiary (Oligocene to Miocene) sediments (Minami 1979). Sediments were deposited in a paralic environment during the initial stage of sedimentation followed by a progressive northward deepening of the basin in early Miocene. This deepening of the basins formed a deep marine facies composed mainly of turbidites. Sedimentation continued into middle Miocene with offlap slope sequences which suggests extensive basin filling. The slope sequences are characterized by numerous paleochannels and mass-flow deposits. In late Miocene the basin was filled again by shallow water sediments. Data on the Korea Strait (still classified since obtained in the early seventies) are expected to show a similar history (see Addendum).

Late Miocene block faulting associated with crustal deformation occurred in the eastern shelf, off northwestern Honshu and in the Korea Strait (Honza et al. 1979; Minami 1979; K.P. Park

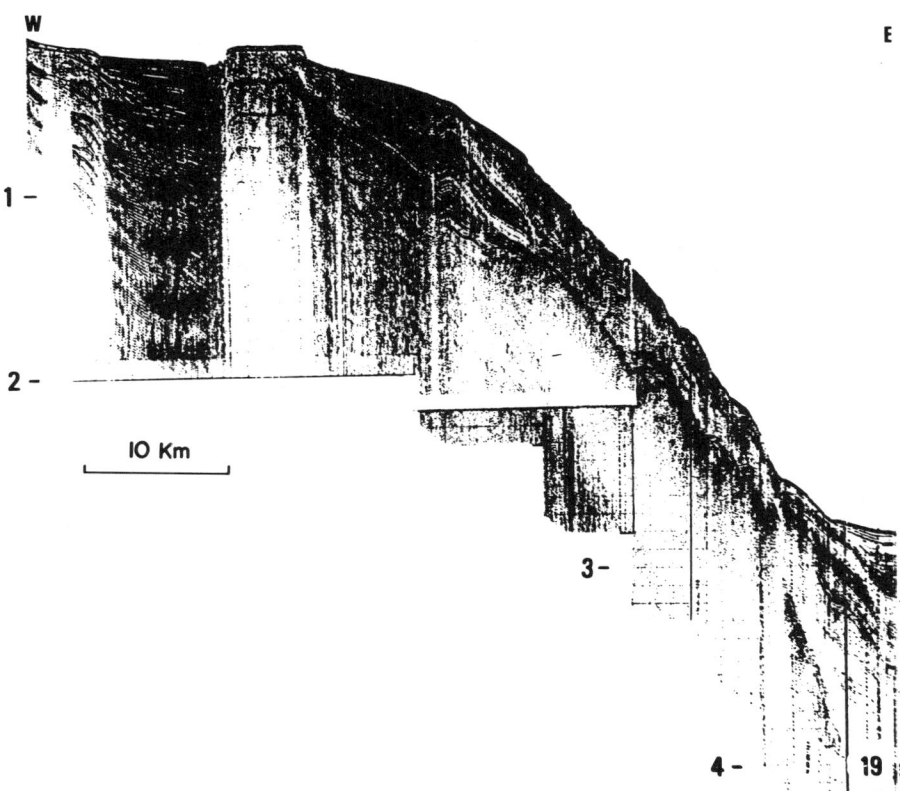

Figure 4.5. *Air-gun profile (line 19 on fig. 4.1) across the eastern margin of the Korean Peninsula showing the fault block that trapped Neogene and Quaternary sediments. Profile courtesy of the Korea Institute of Energy and Resources.*

et al. 1981), during which a pre-Miocene sedimentary sequence was block faulted and folded trending north–south and northeast–southwest (figs. 4.3 and 4.6). The crustal deformation was also associated with volcanism and possible intrusion. Murauchi et al. (1970) suggested that a buried oceanic ridge underlies the northern Honshu shelf. In the Oki Bank and the adjacent shelf area, Miocene sedimentary sequences consist of sandstone, shale, and conglomerate ranging up to 400 m in thickness. The Korea Strait and the southeastern margin of the Ulleung Basin is underlain by up to 3 km of Neogene sediment (Ishiwada and Ogawa 1976) that is folded and faulted, overlying both Paleogene and Cretaceous sediment sequence.

In the East Sea shelf, folded and faulted basement rocks are often exposed on the sea floor. Postdeformational (post-Miocene) sediments are scarce on the inner shelf floor close to the shoreline. They tend to accumulate further offshore forming a series of delta-like prograding sedimentary sequence (fig. 4.7).

Surface Sediments

The distributional pattern of surface sediments on the southern half of the eastern margin of the Korean Peninsula follows, in general, bottom topography because coarse-grained sediment, dominated by sand, muddy sand, and sandy mud, occurs on the shelf (to a depth of about 500 m), whereas the offshore margin is covered with mud (fig. 4.8). In the former, sandy gravel, muddy gravel, gravelly muddy sand, sand, and gravelly mud also occur locally. Exceptions are found on the shelf between Pohang and Ulsan and in the southern embayments where silt and mud prevail.

Recent sediment distribution on the southeastern shelf of the Korean Peninsula appears to be influenced strongly by currents which winnow fine particles into deeper water. The existence of sand and gravel in the middle part of the shelf is suggestive of bottom water flowing from north to south whose effect is diminished south of Pohang. Silts discharged through the Ulsan area appear to be transported largely southward along the coast. The coarse-grained particles on the southeastern corner are shaped most likely by Tsushima Current flowing north– and

Eastern Shelf 85

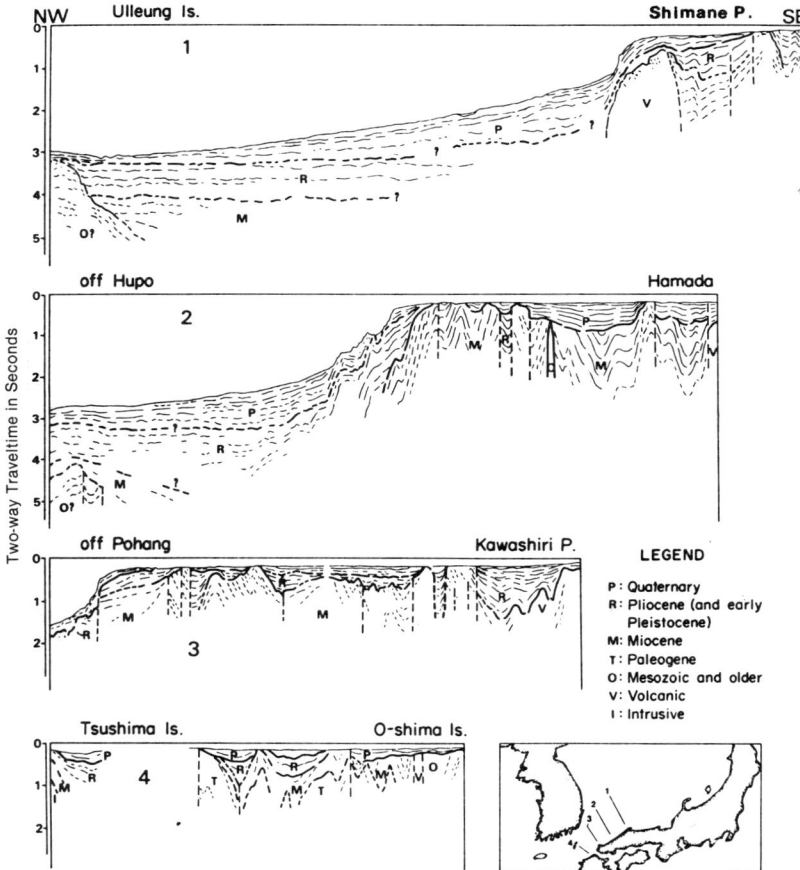

Figure 4.6. *Line drawings of seismic reflection profiles across the East Sea and the Korea Strait. Modified after Honza et al. (1979) by permission of the Geological Survey of Japan.*

86   Marine Geology of Korean Seas

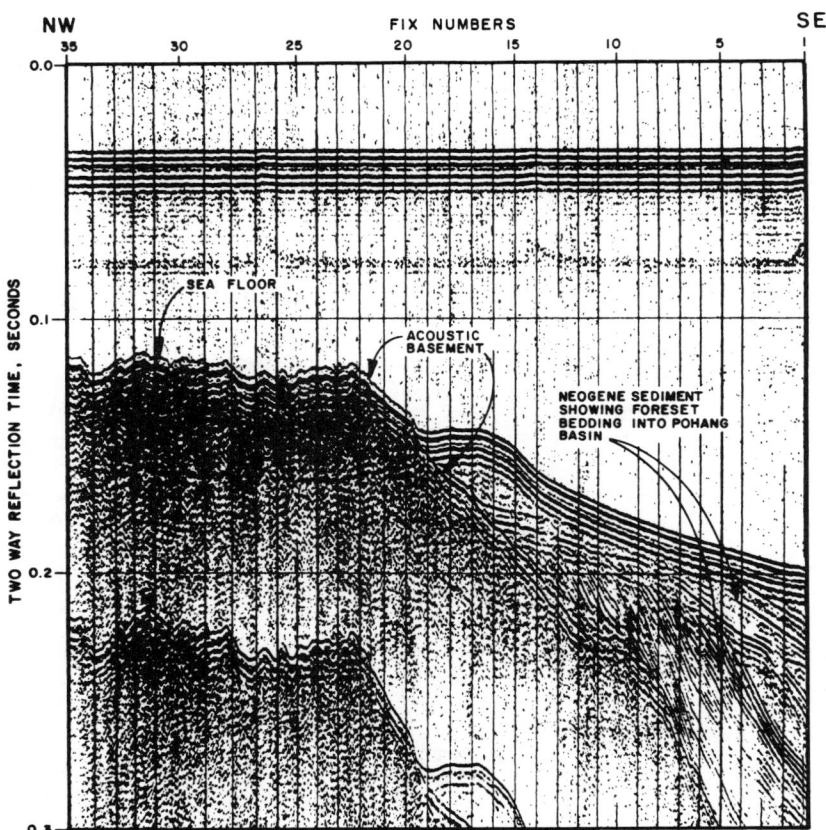

Figure 4.7. Air-gun profile across the eastern shelf of the Korean Peninsula off Bugu (north of Uljin), showing the prograding Neogene and Quaternary sequence. For location see figure 4.1. Figure courtesy of the Korea Institute of Energy and Resources.

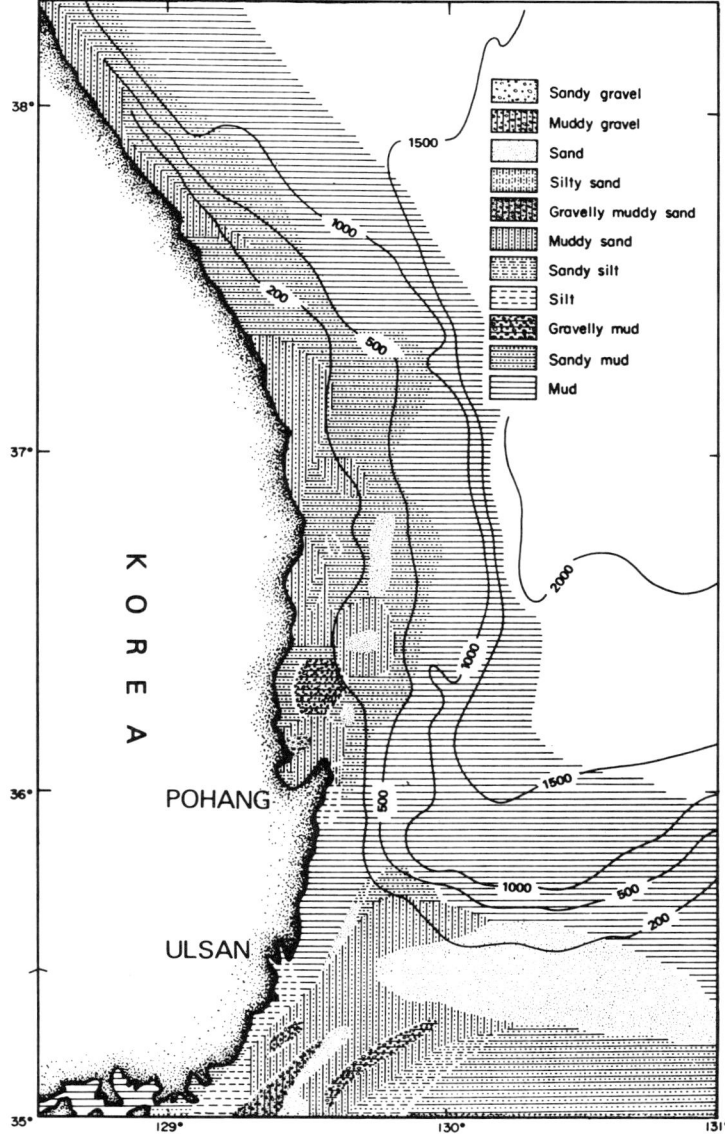

Figure 4.8. *Distribution of surface sediments on the eastern margin of the Korean Peninsula, classified according to the grain size using Folk's (1968) method. Modified after the Chinhae Machine Depot (1979).*

northeast-ward. Sand and gravel in this area are probably relict sediments only partly covered by muddy and silty sediments.

Lack of riverine sediment yield results in a thin cover of sediment on the eastern shelf of Korean Peninsula, often with a rocky sea floor exposed. The bulk of sediments is believed to be produced by cliff erosion and other sources (C.M. Kim et al. 1971). In the Korea Strait off Nagdong River, kaolinites originate largely from the river, whereas illite distribution is influenced by the Tsushima Current (Han 1979). The influence of the Kuroshio in the ecology of foraminifera has also been emphasized by B.K. Kim and Han (1971, 1972) in the southern Korea Strait where warm water species such as *Globorotalia menardii* and *Pulleniatina obliquiloculata* prevail. They also found cold water species such as *Globigerina pachyderma* in the southeastern shelf, signaling cold water infiltration along the coast from the northern East Sea.

5 Ulleung Basin

Physiography

The Ulleung Basin is generally U-shaped with a gentle slope to the northeast (fig. 4.4). The base-of-slope also is transitional to the basin floor. The basin is bordered by the steep continental slope of the Korean Peninsula and the Korea Plateau on the west and north, respectively. The continental shelf on the east coast of Korea is narrow, not more than 25 km wide, and grades to the steep slope (slope gradient, ca 6°). In the south and east, the basin is surrounded by rather gentle slopes of Oki Bank, a submarine extension of Honshu shelf off Shimane Peninsula, and the slope of San-in district of southern Honshu.

The basin extends through the gap between Ulleung and Dok islands to the northeast, forming a narrow and long interbasin plain between the Korea Plateau, Kita-Yamato, and Yamato ridges. It cascades onto the deeper Japan Basin near the southwestern end of Kita-Yamato Ridge. An interplain channel (fig. 5.1) in the Ulleung Interplain Gap runs intermittently through the long axis of the basin connecting the Ulleung Basin to the Japan Basin. The channel, Ulleung Interplain Channel, is approximately 7 km wide and 35 m deep. It is erosional in origin and undercuts the turbidite sequence, formed most likely by strong bottom currents flowing through the gap. A small-scale seamount (named Ulleung Seamount) rises about 700 m above the basin floor approximately 100 km south of the Ulleung Island.

The continental slopes of the eastern Korean Peninsula and the Korea Plateau are steep, characterized by large-scale slump pits and scars (fig. 5.2). On the base-of-slope, they are characterized by the morphology of slump and possible debris flow deposits. Various scarp features also are found on the gentle slope at the foot of the slope in the southern slope and Oki Bank (Ishibashi and Honza 1978).

Acoustic Stratigraphy

According to Tamaki et al. (1978) and Honza et al. (1979), the sedimentary sequences in the basin may be divided acoustically

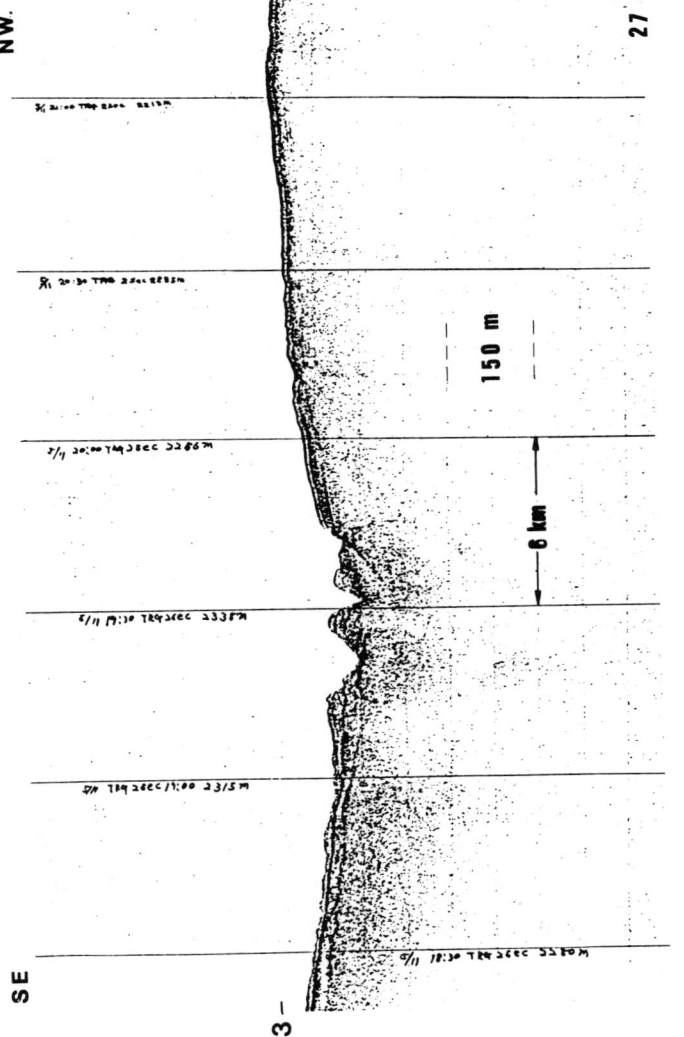

Figure 5.1. 3.5 kHz profile (line 27 on fig. 5.4) across the Ulleung Interplain Gap showing an erosional channel (Ulleung Interplain Channel) incised by strong bottom currents. Vertical scale, two-way traveltime in seconds. Profile courtesy of E. Honza, Geological Survey of Japan.

Ulleung Basin

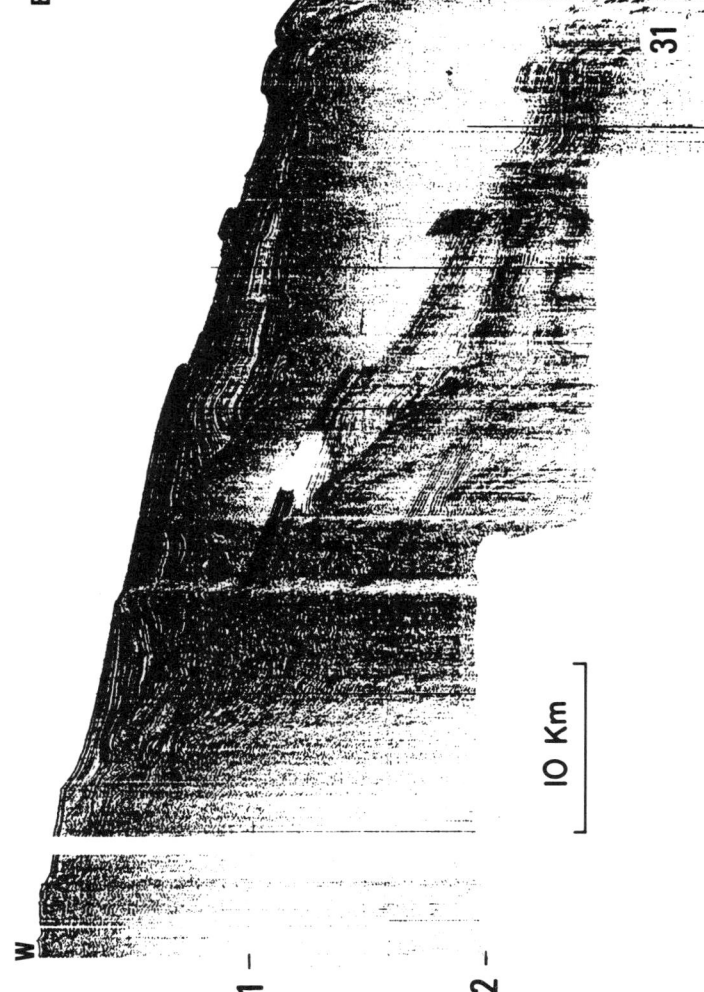

Figure 5.2. Air-gun profile (line 31, for location see fig. 4.1) across the eastern margin (western slope of Ulleung Basin) of the Korean Peninsula showing the slump scars on the slope. Profile courtesy of the Korea Institute of Energy and Resources.

into three units (fig. 4.6). The upper unit (P) is an approximately 0.3 second thick, nondeformed, stratified Quaternary deposit with a high frequency reflection. Underlying this slightly disconformably is the middle unit (R, Pliocene and early Pleistocene in age) which is also stratified and slightly deformed. Near the base-of-slope off San-in, the middle unit attains a maximum thickness of 1.4 seconds. It also increases in thickness southward, continuing in the slope sequences. The lower unit (M) of presumably Miocene age is acoustically transparent attaining more than 1.0 second in thickness.

The top-most sequence of sediment in the basin (about 30–50 m) shows evidence of extensive mass-flow processes during the Quaternary period, associated with a lowstand of sea level (Chough 1982, 1983b).

Mass-Flow Deposits

General Statement

Closely spaced, high resolution 3.5 kHz seismic profiles obtained in the Ulleung Basin reveal various types of mass-flow deposits on the slope, base-of-slope, and basin. The mass-flow deposits recognized on high-resolution profiles include rockfall, slump, slide, grain- and debris-flow deposits, and turbidites with corresponding slide scarps and scars (Jeong 1983; Chough and Jeong, in prep.). These types of deposits have been recognized recently on many continental slopes and related environments both in active and inactive margins (Embley 1976; Jacobi 1976; Nardin et al. 1979; Cook et al. 1982). Variety of mobility is possible in mass flows from rockfall, short-distance rotational slump to mobile debris flow, and turbidity current.

Acoustic Characteristics

Rockfall deposits are acoustically highly reflective and hummocky, and are associated with large-scale, highly peaked hyperbolae imposed on each other (fig. 5.3a). Internal structures are seldom seen and often chaotic due to hyperbolae which are deeply prolonged into the mass. Side echoes are common on the surface.

Submarine slides are recognized generally on the profiles by minimal translation of slide blocks of mass that are internally less deformed than the slumps and retain their original bedding. Slump deposits with minimal translation are recognized seismically by contorted internal beds, by discontinuous internal reflectors, and/or blocky surface morphology (fig. 5.3b). When the bedding is not always present, the distinction between the two becomes difficult. The appearance of the displaced material on a seismic record also depends on the degree of lithification of the material. If the material is unconsolidated mud, even a short displacement can disrupt the continuity of internal reflectors sufficiently to make the material appear acoustically transparent.

Debris-flow deposits appear usually transparent due to poor sorting or lack of distinctive internal structure (fig. 5.3c), similar to transparent hemipelagic or pelagic sediments. However, the former is usually lens-shaped and internally discontinuous. The surface morphology and reflectivity of mass-flow deposits vary from hummocky to relatively smooth. In some cases, a relatively distinct echo is present whereas, in others, a diffuse or prolonged echo is observed. The irregularity produced during the flow or blocks of semilithified sediments has been observed on the surface on many sonar records (Embley 1976).

Turbidites are acoustically distinctive due to the internal layers or bedding. The beds are emphasized by an alternation of coarse and fine sediments.

Zoned Facies

Mass-flow deposits in the Ulleung Basin and its slope are zoned in a contour-parallel fashion because rockfall, slump, and slide deposits occur mainly on the slope which are, in turn, replaced in the deeper area below slope by debris flow deposits (fig. 5.4). Debris flow deposits are transitional to the turbidites in the central basin. This zoned facies relationship points to compelling evidence that mass flows were generated from a line source rather than from a point source. The lack of submarine canyon and deep-sea fans on the surrounding slope and base-of-slope also supports the uniform sheet-like nature of turbidity currents. High resolution profiles on the surrounding shelf also show a prograding toplap sequence (fig. 5.3c). This suggests an active transport of sediments toward the shelf break.

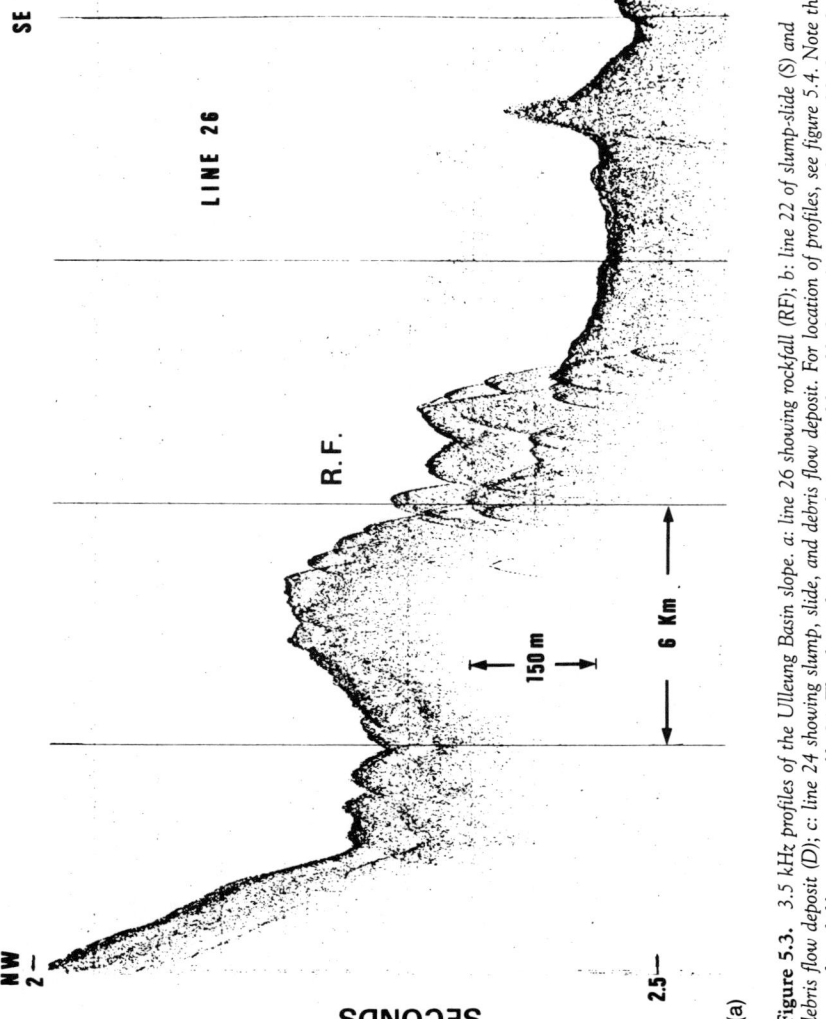

Figure 5.3. 3.5 kHz profiles of the Ulleung Basin slope. a: line 26 showing rockfall (RF); b: line 22 of slump-slide (S) and debris flow deposit (D); c: line 24 showing slump, slide, and debris flow deposit. For location of profiles, see figure 5.4. Note the prograding shelf sequence on profile 24. The slump deposits are characterized by a blocky surface, contorted internal beds, or by discontinuous internal reflectors. Debris flow deposits are recognized by the lack of reflectors and by the acoustically transparent lens-shape configuration. Profiles courtesy of E. Honza, Geological Survey of Japan.

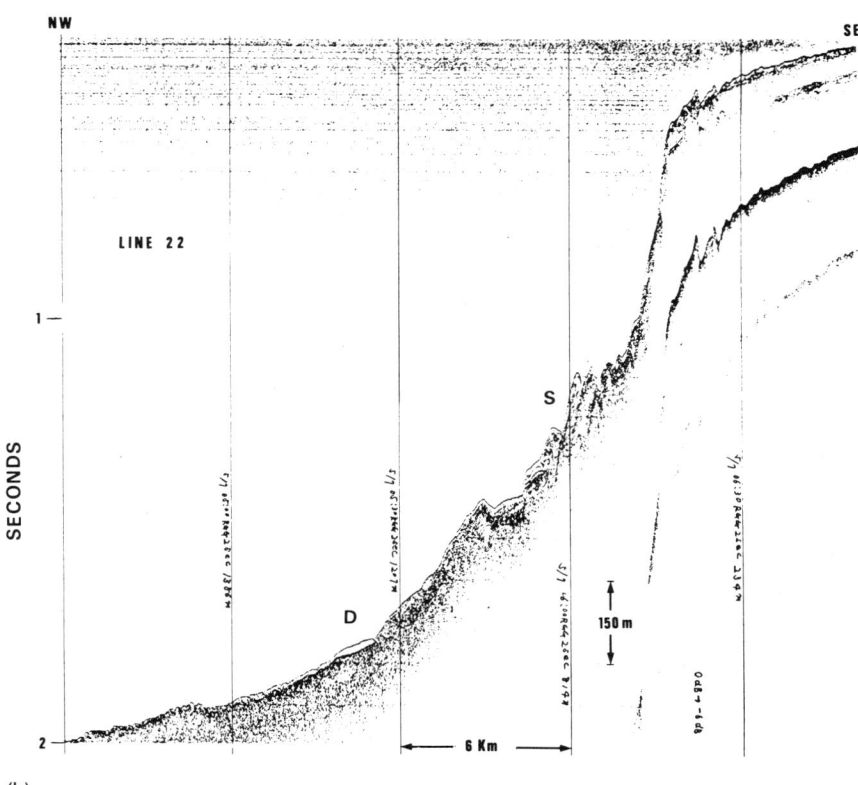

Figure 5.3b.

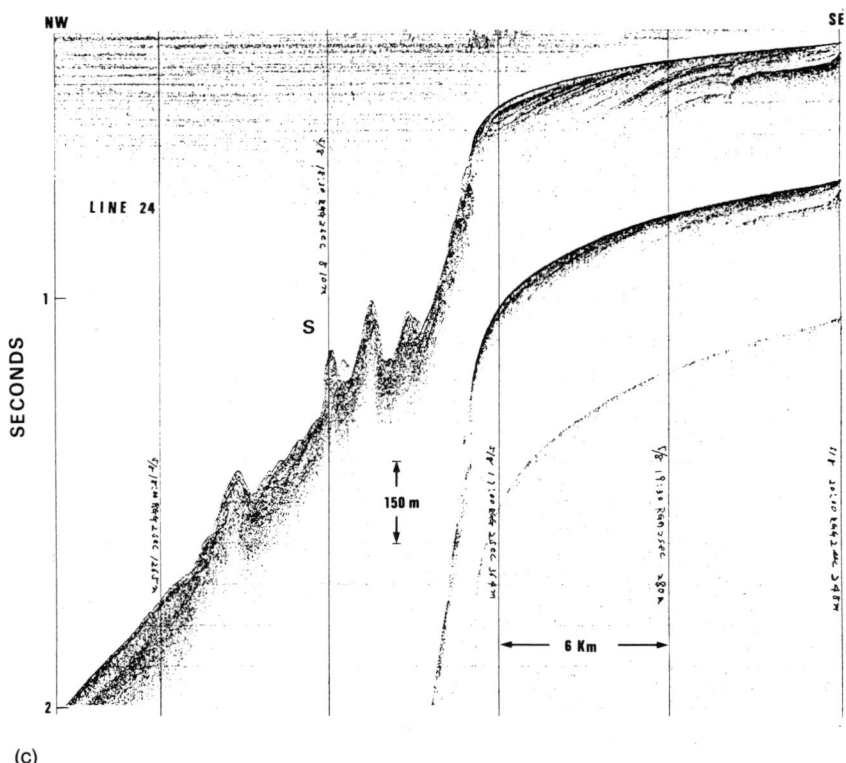

Figure 5.3c.

Turbidite Facies

General Statement
Since about late Pliocene time, the accumulation of turbidites has been predominant over pelagic sedimentation in the deep basins of the sea. The evidence of mass-flow processes on the slopes of the Ulleung Basin corresponds to the existence of thick, layered sediment sequences (Tamaki et al. 1978), dominantly of turbidite origin (Chough 1982, 1983b; G.H. Lee 1983). The submarine mass-flow processes have been more common probably during glacial periods when sea level was lower, transporting terrigenous debris directly to the outer continental shelf via the adjoining rivers and streams.

Thick turbidite sequences dominated by finely laminated and homogeneous muds alternate with nonturbiditic muds (fig. 5.5). The latter include indistinct laminae mud deposited probably by bottom currents and bioturbated hemipelagic mud. The finely laminated and homogeneous turbidite mud facies are associated with a minor occurrence of massive or graded gravel and sand layers, micro-cross-laminated sands, and silts. The hemipelagic facies are confined mainly in the upper 2 m of the cores, whereas the turbidites are predominant below it. Some fine-grained turbidite layers consist also of abundant biogenic sediments such as diatoms, silicoflagellates, spicules, radiolarians, and others, similar to those of hemipelagic facies.

Coarse-Grained Turbidites
Coarse-grained deposits, layers of sand and gravel, were found in the Ulleung Interplain Gap. Layers are up to 30 cm thick and show distinct graded or massive units bounded at the base by a scour surface and overlain by parallel-laminated sand division. Gravels are up to 12 mm long in diameter including large amounts of resedimented volcanic debris. The parallel or crudely parallel-laminated sand division (T_b) occur usually overlying the A-division. They are composed of terrigenous, volcanic and, in some cases, foraminiferal sands.

The microripple cross-laminiated and climbing microripple-laminated sand and silt occur rarely exceeding 1 cm in thickness. The convolute mud unit also occurs and consists of slumped and sheared interval of about 20 cm in thickness.

98   Marine Geology of Korean Seas

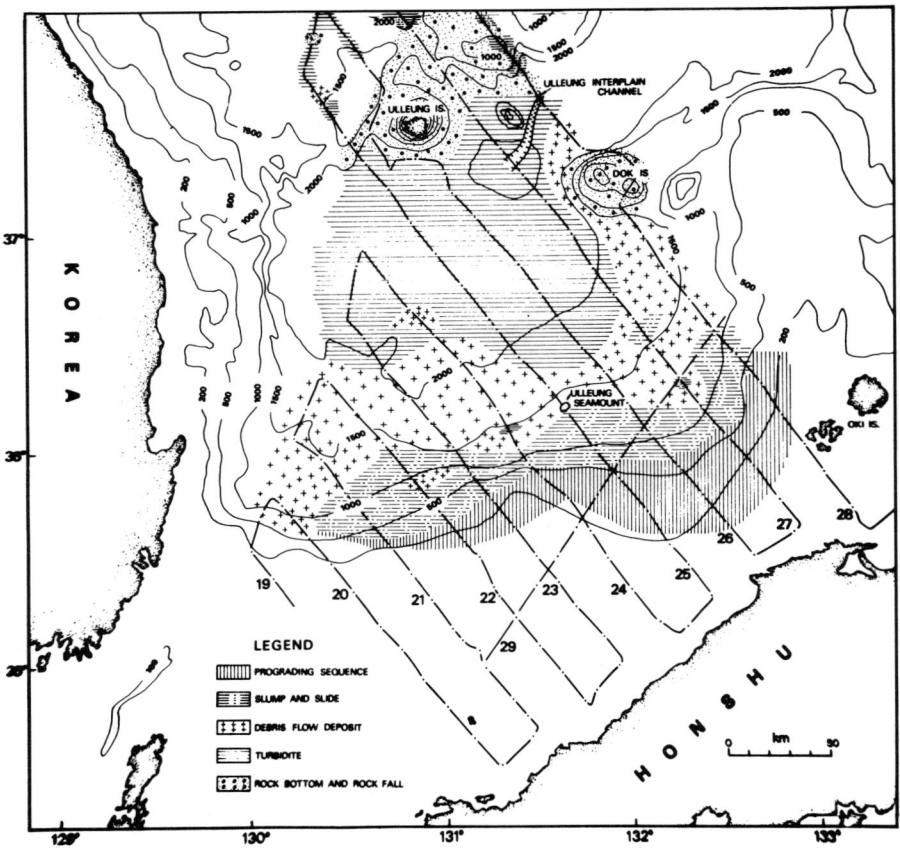

Figure 5.4. Facies distribution of mass-flow deposits in the Ulleung Basin and its margin. After Jeong (1983) and Chough (1983b) by permission of the Geological Society of London.

Figure 5.5. Core descriptions (Ulleung Basin, for core locations see fig. 5.7); asterisk for the depth of a C-14 age (18,000 ± 640 y B.P.). The top tephra (ash) layer (circled number, 1) is called the Ulleung ash (Oki ash) erupted from the Ulleung Island at about 9300 y B.P.; the second layer is the Aira-Tn ash (circled number, 2) erupted from the Aira Caldera in southern Kyushu at about 21,000–22,000 y B.P.; the third layer (circled number, 3) is probably either the Yamato ash (25,000–35,000 y B.P.) or the Aso-4 ash from Aso Caldera (50,000 y B.P.) (Arai et al. 1981). Numbers on the right side of each column indicate the depth points of subsampled turbidite and hemipelagic layers. Modified after Bahk and Chough (1983) by permission of the Journal of Sedimentary Petrology.

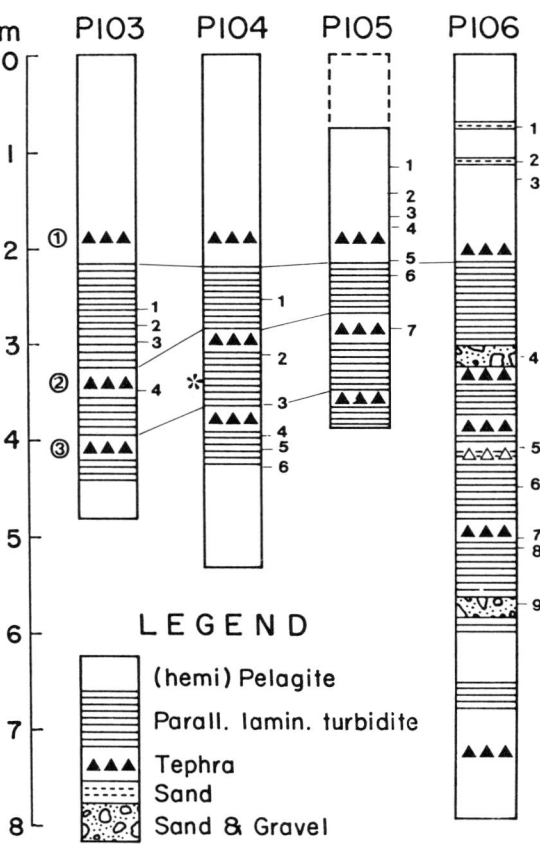

Turbidite Muds
Parallel-laminated mud, designated E_1 following Piper's scheme (1978), is the dominant unit in the turbidites of the Ulleung Basin (fig. 5.6a). It consists of various colors of olive grey and grey olive (5 Y 5/2–5 Y 3/2) mud. In most cases the unit repeats itself, although it occasionally overlies the microripple or cross-laminated division. The individual depositional units are ill-defined (fig. 5.6b) and rarely exceed 1 cm in thickness. Sand fractions in the laminated muds are composed mainly of foraminiferal grains of planktonic origin. They are concentrated near the base of each unit with discrete lenses of silt and linear particles oriented parallel to the bedding. Both carbonate and organic carbon contents in the laminated muds range up to an average of 24% and 1.2%, respectively.

The top of the parallel-laminated unit is designated as homogeneous mud (designated E_3). In X-radiographs this division is devoid of parallel laminae and thereby differentiated from the underlying laminated division. The thickness of each unit ranges up to 4 cm and consists of terrigenous as well as biogenic ooze components. The homogeneous mud is characterized by a near-absence of coarser than 30 μm fractions and is thus well sorted.

Hemipelagic Facies

The present sea floor of the Ulleung Basin is covered with hemipelagic sediments rich in diatoms and fine-grained terrigenous materials with minor amounts of silicoflagellates, spicules, and others. The sediments are bioturbated largely by benthonic deposit feeders and contain abundant pyritized filaments of similar origin (fig. 5.6c). In the sediment column, materials retain the evidence of near-bottom current activity or incorporation into dilute turbid suspensions, referred to as nepheloid layers. Other than the pyritized filaments, burrows rarely occur in the pelagic sediments and in the topmost sections of each turbidite layers. Chondrites and rind burrows are found commonly. In many sections the sediments are completely mottled and disrupted by organisms.

The indistinct laminae muds (fig. 5.6d) occur commonly in the Ulleung Interplain Gap where the existence of bottom currents

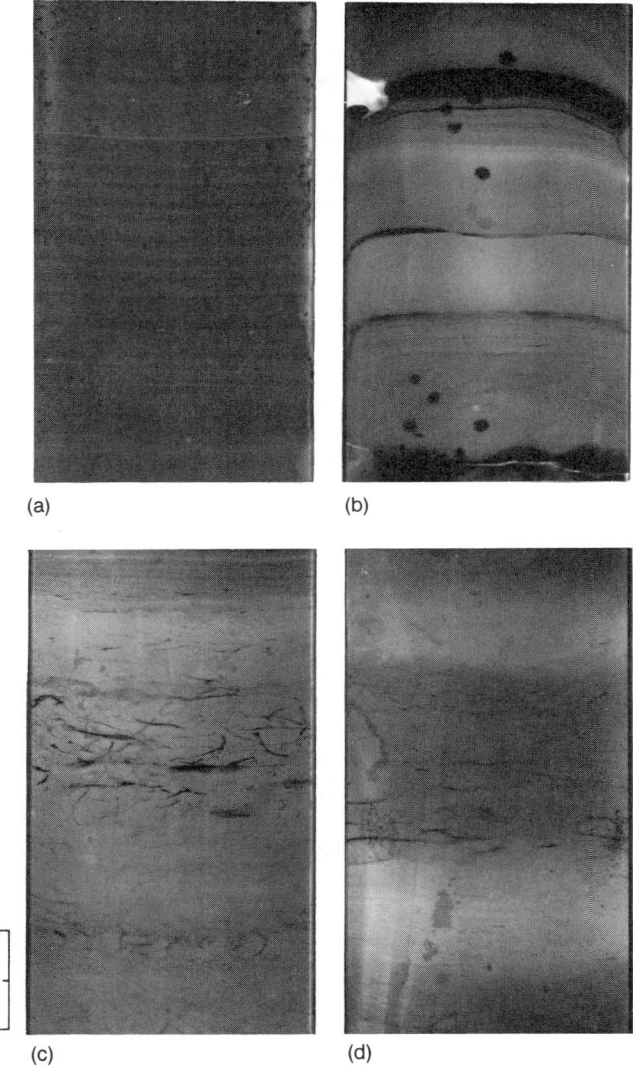

Figure 5.6. X-radiographs of sediment cores from the Ulleung Basin. a: core P103, core depth, 305–315 cm; b: core P105, core depth, 162-172 cm; c: core P103, core depth 442–452 cm; d: core P103, core depth, 389– 399 cm. For core locations see figure 5.7. In (a) the repetition of parallel-laminated mud (E_1) is mostly ill defined. Homogeneous turbidite mud (E_3) in (b) overlies either the micro-cross-laminated silt or the parallel-laminated mud. Burrow-rich hemipelagic mud layer in (c) and (d) alternates with rather homogeneous and less-burrowed mud layer.

is suggested on the high resolution seismic profiles. The muds consist of light olive grey (5 Y 5/2, 5 Y 5/1) mud alternating with slightly darker olive grey (5 Y 3/2) mud (fig. 5.6d). In the latter, the sediments consist mostly of biogenic material rich in diatoms, are largely bioturbated, and contain some pyritized filaments.

Stratigraphy

The abundant ash layers in the cores provide a tool for the correlation of individual turbidite and hemipelagic layers. The composition of each layer and the time of eruption have been elaborated for the entire region by Arai et al. (1981). Layers identified in the Ulleung Basin cores include the Ulleung ash. This is composed primarily of colorless pumiceous glass in the top hemipelagic layer of about 180-190 cm that was erupted from the Ulleung Island at about 9300 y B.P. (fig. 5.5) (Arai et al. 1981). The phenocrysts in the ash include alkali feldspar and hornblende.

Another prominent ash layer is the Aira-Tn ash erupted from the Aira Caldera in southern Kyushu at about 21,000-22,000 y B.P. This vitric ash consists mainly of bubble-walled and pumiceous glass, which is correlated for the entire basin occurring at about 300 or 345 cm. A C^{14} age determined at the depth of 360-364 cm (core P103) gives an age of 18,000 ± 640 y B.P. This is slightly younger than that postulated by the ash correlation (fig. 5.5).

Other ash layers underlying the Aira-Tn ash include those of probable Yamato ash (25,000-35,000 y B.P.) from Ulleung Island and Aso-4 ash from Aso Caldera (50,000 y B.P.) (Arai et al. 1981). In core P106, a number of ash layers whose individual age is not certain were found below it. In addition to the ash particles pelagically settled, there are a few ash layers composed mainly of ash debris that were resedimented by turbidity currents. The correlation of ash layers and lithologic changes in the cores yield the sedimentation rate of pre-Holocene turbidites of 12 cm/10^3 y. The average thickness of individual turbidite layers is less than about 6 mm. This gives a minimum recurrence time of about 50 years, assuming that the individual layer was deposited in a relatively short period (days or months) of time.

Turbidite Provenance

General Statement
The provenance of turbidites in the Ulleung Basin was deciphered in detail by Chough et al. (1981b), Bahk (1982), and Bahk and Chough (1983). They determined the relative amounts of heavy minerals found on the margin of the basin and compared statistically these amounts with those in the turbidite layers in cores from the basin. This was possible because the basin is bordered by the continental margin of eastern Korea, Korea Plateau and Oki Spur, and the northwestern Honshu, in which rocks are of different assemblage and hence represent different detrital mineral provinces. This permits evaluation of the relative importance of tectonics and sedimentation in the back-arc area in terms of origin, volume, and rate of sediment yield. In the Japan Basin, turbidites were derived mainly from the Asian continent, whereas those in the Yamato Basin were derived mainly from the Japanese islands (Sibley and Pentony 1978).

Geologic Setting
The east coast of Korea lies in the eastern slope of north–south trending Taebaeg mountain range whose slope is relativley steep and is drained by numerous short streams and creeks (e.g., Sang, Yungog, Hosan, Bugu, and Song streams) (fig. 5.7). The drainage area of the Sang and Yungog streams consist primarily of Jurassic granite and the Precambrian Yeoncheon System and its associated schist and gneiss. The rocks in the drainage areas of Hosan and Bugu streams are characterized by a stable craton of Precambrian schist and granitic gneiss as well as local occurrence of Jurassic granite. Along the southern coast lies the Cretaceous sedimentary basin of the Gyeongsang Supergroup, east of which the Pohang Tertiary basalt and Cretaceous porphyrite are exposed. The Ulleung and Dok islands in the north of the Ulleung Basin consist of Pliocene and early Pleistocene alkali volcanic and trachytic rocks.

The Oki Spur is a northern extension of the Shimane Peninsula and connected to the Oki Bank on the northeast. It separates the Ulleung Basin in the west from the Yamato Basin in the east. The Oki Islands on the spur consist mainly of

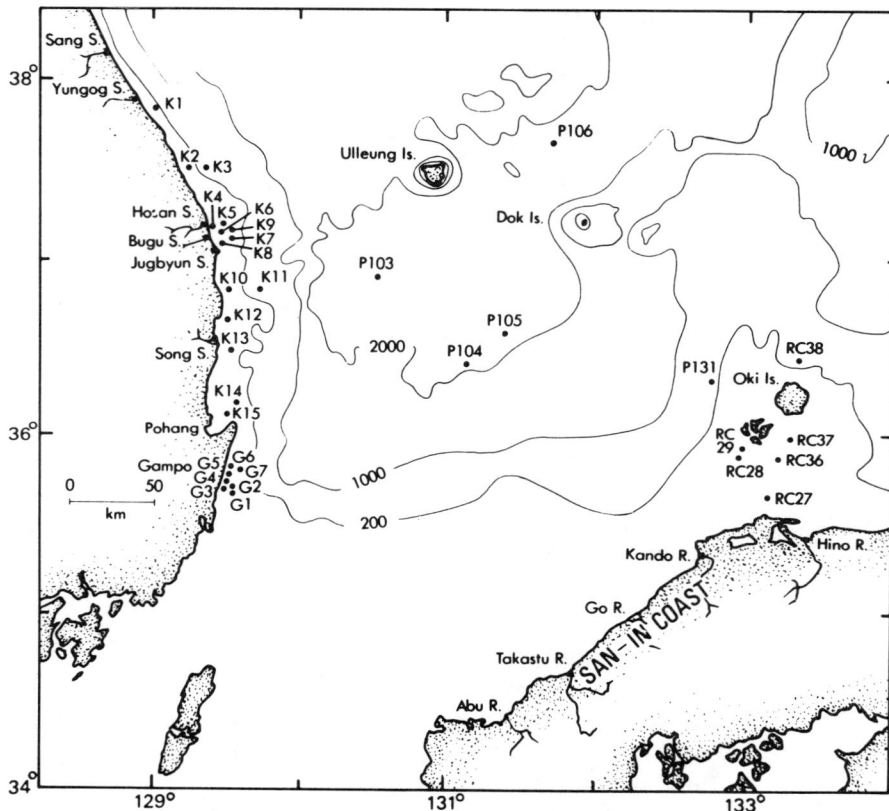

Figure 5.7. Location of core and grab samples taken from the Ulleung Basin and the adjoining margins. P: piston core (R/V Hakurei-maru); RC: rock corer; others are grab samples from the shelves, rivers (R), others and streams (S). Contours in meters. After Bahk and Chough (1983) by permission of the Journal of Sedimentary Petrology.

Pliocene and Pleistocene alkali volcanic and trachytic rocks covering or intruding the Miocene sedimentary, pyroclastic, and andesitic rocks. Precambrian gneiss is locally exposed in the eastern part of the Oki Islands. On the other hand, the Shimane Peninsula is mostly composed of Miocene sedimentary rocks intercalated with rhyolitic and andesitic rocks. In the region south of the peninsula, Cretaceous granitic rocks are widely distributed. The drainage area of Hino River consists of Neogene hornblende andesite, basalt, granite, and older (Precambrian to early Triassic) crystalline schist and phyllite. Rocks in the southwestern San-in coast include Tertiary andesite, dacite and basalt, late Cretaceous rhyolite, dacite and granite, and older schist and limestone.

Heavy Mineral Abundance
The total amounts of heavy minerals in the 63–125 μm size fraction range from 0.4 to 5.2% and from 1.6 to 8.3% in the continental shelf of Korea and in the Oki Spur of Honshu, respectively. Stream samples contain larger amounts (up to 35.2%). In the Ulleung Basin, turbidite layers contain commonly 2.4% (calculated from 26 samples) of heavy minerals in the 63–125 μm size fractions. The heavy mineral abundance in the Ulleung Basin is comparable to those in modern sea-floor sands from different tectonic settings (Dickinson and Valloni 1980).

In the samples from the western margin of the basin (continental shelf of Korea), the heavy mineral assemblage is dominated by green hornblende and stable minerals of metamorphic origin such as garnet, epidote, sillimanite, staurolite, and actinolite-tremolite (fig. 5.8a). Sillimanite and staurolite are restricted in the western margin of the basin and are practically absent in the rest. Garnet and brownish tourmaline grains also are abundant in the western margin (av. 13.9 and 2.5%, respectively). In the eastern margin of the basin (Oki Spur and San-in Coast of Honshu), green hornblende is also predominant but followed by unstable minerals such as orthopyroxene and clinopyroxene and basaltic hornblende with minor amounts of ferrohastingsite, epidote, brown hornblende, and olivine (fig. 5.8b). Basaltic hornblende (up to 47%) and ferrohastingsite (up to 26%) are the most diagnostic minerals in this region. Clinopyroxene is mainly augite showing abundant saw-teeth marks on the crystal edge or

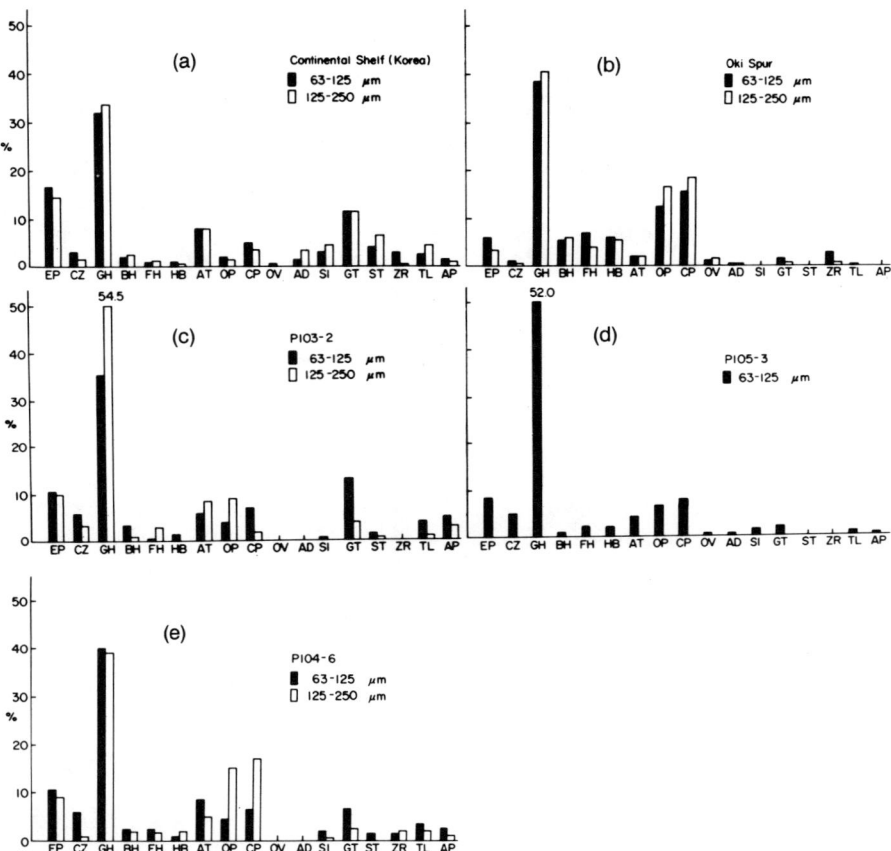

Figure 5.8. Histograms showing average frequency of heavy minerals from the continental shelf of Korea (a) and from the Oki Spur (b). A representative turbidite layer [P103-2; (c)] shows a similar assemblage to (a), whereas another, P105-3 (d), corresponds to (b). The P104-6 (e) sample contains both assemblages of (a) and (b): EP = epidote; CZ = clinozoisite + zoisite; GH = green hornblende; BH = brown hornblende; FH = ferrohastingsite; HB = basaltic hornblende; AT = actinolite + tremolite; OP = orthopyroxene; CP = clinopyroxene; OV = olivine; AD = andalusite; SI = sillimanite; GT = garnet; ST = staurolite; ZR = zircon; TL = tourmaline; AP = apatite. After Bahk and Chough (1983) by permission of the Journal of Sedimentary Petrology.

an euhedral shape with occasional glass bubbles. Most orthopyroxene grains present are hypersthene and are predominantly euhedral with glassy rim and glass bubbles of a volcanic origin. Olivine, a rare detrital mineral on other continental shelves, occurs in considerable amounts in Oki Spur ranging up to 6%.

Heavy mineral assemblages in the Ulleung Basin are dominated by green hornblende with various mixtures of stable and unstable minerals in different turbidite layers. Some layers contain either the assemblage much similar to those on the western margin (fig. 5.8c) or that on the other margin (fig. 5.8d). Other layers contain a mineral assemblage common to both margins (fig. 5.8e). This is demonstrated further by using Q-mode factor analysis.

Factor Analysis

The results of factor analysis show that four factors account for 90.1% of the variance or sample interrelationships. Factor I is represented by garnet, actinolite-tremolite, zircon, sillimanite, and staurolite accounting for 29.2% of the total sample variance. Factor II is heavily weighted on orthopyroxene, basaltic hornblende, and ferrohastingsite totaling 22.8% of the sample variance. Factor III is represented by epidote and clinozoisite (25.4%). Factor IV is heavily weighted on clinopyroxene, brown hornblende, and apatite (12.7%).

Provenance

Factors I-IV are compared with each other to diagnose the interdependence of each contributing factor. In figure 5.9, the samples from the east coast and continental shelf of Korea (represented by Factor I) are shown in good contrast to those from the San-in coast and Oki Spur (represented by Factor II). The turbidite layers can be divided into three groups. The first group includes P103-1, 2, 3; P104-1, 3, 4, 5; and P105-6 and is close to the Factor I axis. This suggests that the first group was derived mainly from the stable craton of the east Korea continental margin. Most of these turbidite layers (such as P103-1, 2, 3; P104-1, 5; P105-6) were deposited between major volcanic eruptions. P104-4 and 5, deposited prior to the third tephra layer, also were derived from the eastern continental margin of

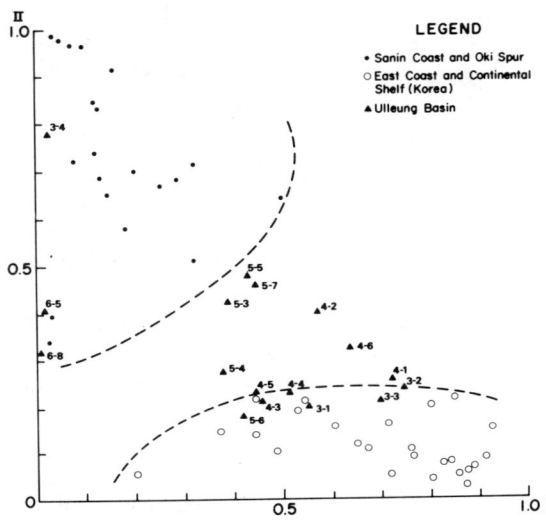

Figure 5.9. Factor loadings for the heavy mineral data. Factor I, heavily weighted on garnet, actinolite-tremolite, sillimanite, staurolite, and zircon, represents the samples from the east coast and continental shelf of Korea. Factor II of orothopyroxene, basaltic hornblende, and ferro-hastingsite is due to contributions from the Oki Spur and San-in coast. The first number of Ulleung Basin samples is the last digit of core, e.g., P103-3. The second number is for core depth shown in figure 5.5. After Bahk and Chough (1983) by permission of the Journal of Sedimentary Petrology.

Korea. The second group (P103-4, P105-3, 5, 7, and the turbidite layers of P106) falls in or near the Factor II axis. This suggests a derivation from the San-in coast and Oki Spur. P103-4 and P105-7 were deposited nearly coincident with the eruption of Aira Caldera. Two representative layers (P106-5 and 8) of the core from the Ulleung Interplain Gap consist of a different mineral assemblage from the others, suggesting that the turbidites of P106 were derived from the adjoining volcanic islands and the Oki Spur. Other turbidite layers (third group), such as P104-2 and P105-4, belong to the region of either margins. These layers contain mineral assemblages common to both margins. Their provenance could only be resolved, if not impossible, by a further analysis of sedimentary structures and other factors.

6 Coastal Embayments (South Coast)

General Statement

The southern coast of the Korean Peninsula is characterized by numerous postglacial embayments and nearshore islands forming a typical ria-type coast. Sedimentation is controlled largely by tidal currents depositing fine-grained sediments. These sediments are either riverborne or transported from offshore, such as in the Jinhae Bay (B.K. Park et al. 1976). Only a few large rivers drain into the southern coast, such as the Yeongsan River on the southwestern coast and Nagdong River on the southeastern coast. These rivers deliver a substantial volume of clastic sediments, forming estuarine environments. The Nagdong River discharges approximately 63 million tons and delivers about 10 million tons of sediment into the sea annually (Ministry of Construction 1974). The major portion of discharge (about 71%) occurs during the summer floods. The Nagdong River drains about 24,000 km^2 in area which is composed largely of Cretaceous sedimentary rocks and granites (Gyeongsang Basin). Sand and silty sand prevail in the upstream and mouth of the river, whereas finer-grained sediments are dominant seaward (W.H. Kim and Park 1981).

Other embayments (fig. 6.1) with limited drainage systems have recently been studied in detail, such as the Gamagyang Bay (H.J. Kang 1981; H.J. Kang and Chough 1982), the Deugryang Bay (J.H. Chang et al. 1980), and the Jinhae Bay (B.K. Park et al. 1976; Hahn and Kim 1977). These studies provide a model on the processes of coastal embayments on the South Sea dominated by tidal currents with no significant sediment yield from the surrounding drainage area. The sedimentary processes in the Gamagyang and Deugryang bays are outlined next.

Gamagyang Bay

Physiography
Gamagyang Bay (150 km^2 in area) occurs on the ria-type southern coast of the Korean Peninsula (fig. 6.1). The bay is bounded on the north by relatively high mountains and hills which are steep and rise more than 400 m above sea level. The bay is connected to the South Sea by numerous tidal inlets be-

Figure 6.1. Index map of Gamagyang Bay (D.B.: Deugryang Bay; J.B.: Jinhae Bay; G.B.: Gwangyang Bay; G.I.: Geoje Island; N.I.: Namhae Island). After H.J. Kang and Chough (1982) by permission of Marine Geology, Elsevier Scientific Publishing Co.

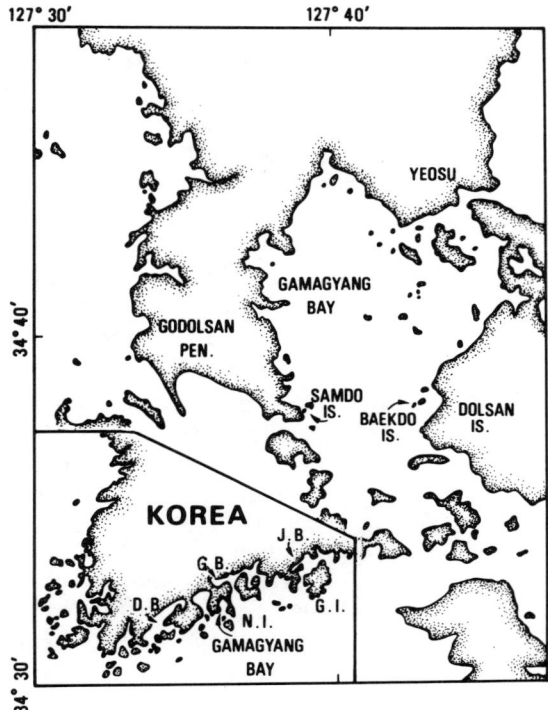

tween relatively low islands, whereas a tidal inlet connects the bay to the Gwangyang Bay and open sea in the northeast (fig. 6.1). The Gamagyang Bay region is characterized by the late Cretaceous rhyolitic tuffs. The drainage basin of the bay is limited, bounded by small streams and creeks through which sediment discharge is minimal.

The bay is shallow (mean water depth, about 9 m) and characterized by a semidiurnal tide flowing largely north and south (fig. 6.2). The maximum spring tidal range is up to 350 cm with a mean of 195 cm (National Hydrographic Office of Korea 1980). In the southern inlets, the surface tidal currents attain a maximum velocity of about 2.5 knots (National Hydrographic Office of Korea 1973).

Sea Floor

The bathymetric chart (fig. 6.3) shows generally a monotonous and flat sea floor except for the moats and depressions in the vicinity of islands and points. Sediment waves occur near the mouth of the northeastern inlet. They are about 30 to 50 m in wavelength and about 30 to 50 cm in amplitude. The waves also occur at a depth of 7 to 8 m below the sea floor. Mean grain size of the sediments comprising the waves is about 7.5 ϕ and slightly coarser than other sediments on the adjacent sea floor. The latter are mostly composed of silty clay (fig. 6.4).

The depressions or moats near the islands and points are ubiquitous, elongated in the north–south direction of flood and ebb currents. They are usually 15 to 25 m deep and more than 500 m across (fig. 6.5). In cross section they are U-shaped, whereas elliptical in a longitudinal view. The low reflectivity on the bottom of each moat suggests that the moat is covered with fine-grained sediments. A gradual decrease of mean grain size toward the center is observed in the moats (fig. 6.5).

Acoustic Stratigraphy

The acoustic basement deepens progressively southward with two paleovalleys (500 to 1000 m wide and 5 to 10 m deep) running both north- and northeast-ward. The thickness of the sediment sequence above the acoustic basement ranges from a few meters to more than 30 m (fig. 6.6). Thick accumulation occurs in the central part of the bay.

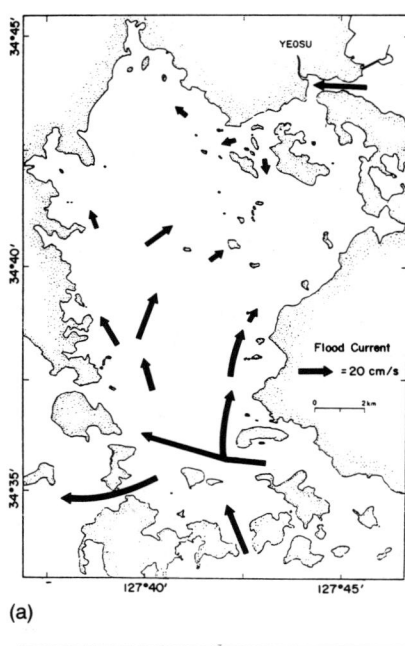

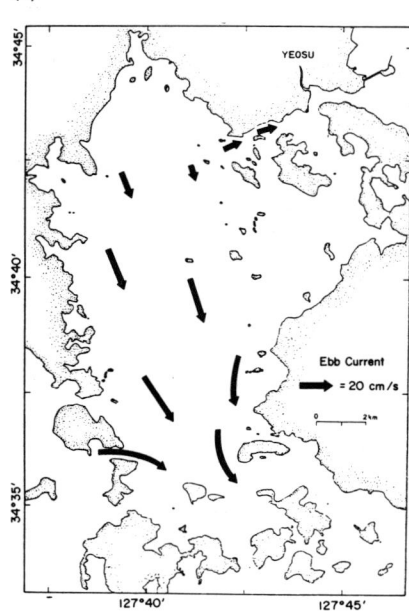

Figure 6.2. Tidal currents in the Gamagyang Bay. Arrow length indicates average velocity of a series of observations. a: flood current; b: ebb current. After Chough et al. (1982) courtesy of the Korea Ocean Research and Development Institute Bulletin.

Figure 6.3. Bathymetry of Gamagyang Bay, corrected in mean sea level. Depth variation due to temperature change was considered negligible. Contours in meters. After H.J. Kang and Chough (1982) by permission of Marine Geology, Elsevier Scientific Publishing Co.

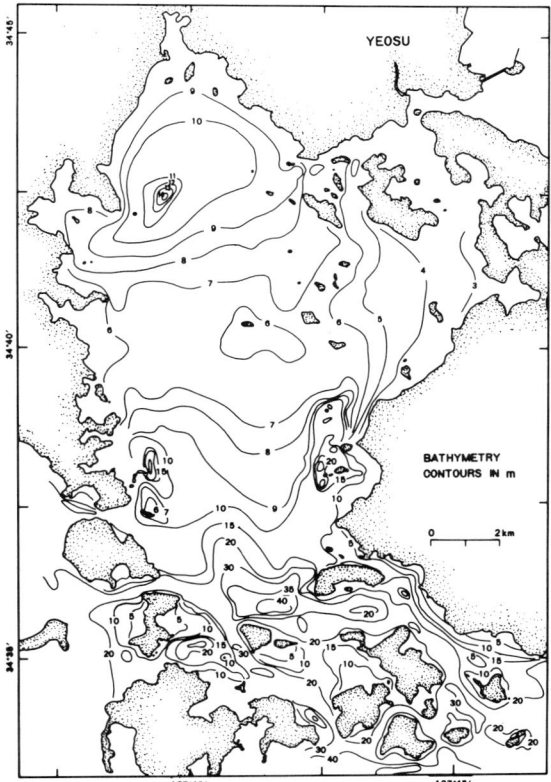

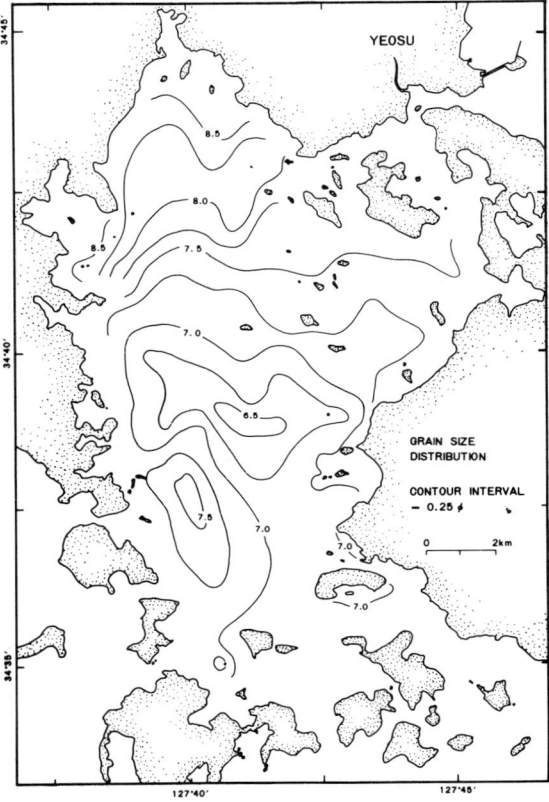

Figure 6.4. Grain-size distribution of more than 70 surface samples showing a progressive northward decrease of size. After H.J. Kang and Chough (1982) by permission of Marine Geology, Elsevier Scientific Publishing Co.

Coastal Embayments (South Coast) 117

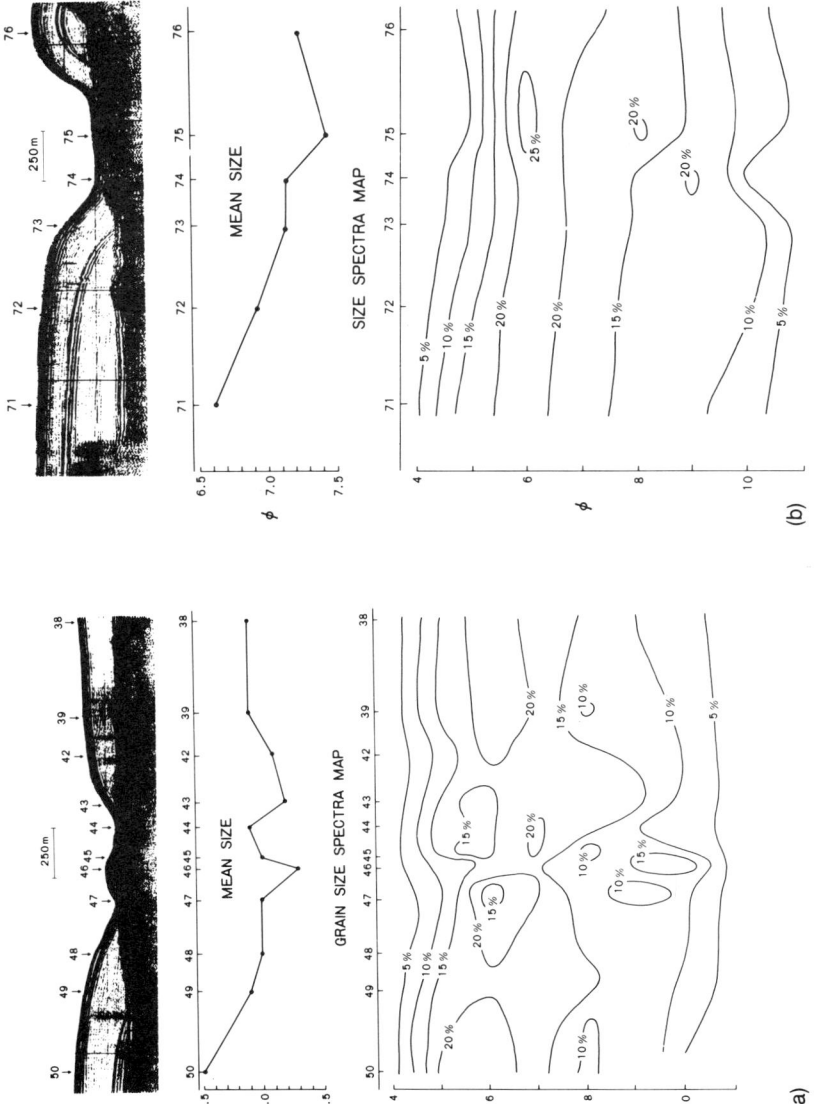

Figure 6.5. Mean grain-size variation and grain-size spectra map of a series of surface samples taken from the moat. a: west of Baekdo Is.; b: east of Samdo Is. For location see figure 6.1. After H.J. Kang and Chough (1982) by permission of Marine Geology, Elsevier Scientific Publishing Co.

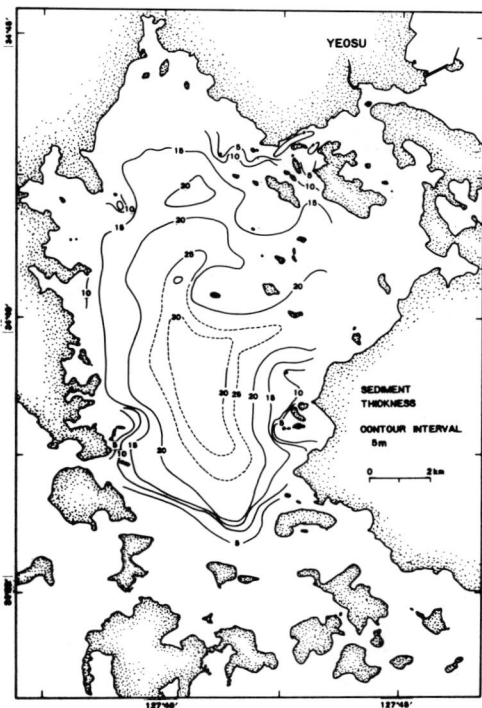

Figure 6.6. *The thickness of sediments above the acoustic basement. Assumed sound velocity is 1600 m/s in the sediments. The dotted lines indicate the estimated thickness, whose layers are masked by the turbid layer caused by the existence of probable gas bubbles in the sediments. After H.J. Kang and Chough (1982) by permission of Marine Geology, Elsevier Scientific Publishing Co.*

Figure 6.7 shows the existence of a relatively strong mid-reflector within the sedimentary sequence. The mid-reflector, characterized by a relatively large contrast in acoustic impedance, occurs at the depth of about 15 to 20 m below the sea floor and at a progressively shallower depth toward the margin. It can be traced over the entire bay except for the local high of the acoustic basement.

According to the origin and nature of sedimentation, the sedimentary sequence in the moats or depressions may be classified as either "depositional" or "depositional-erosional." The former is characterized by laterally continuous and concordant reflectors, whereas the latter exhibits a depositional phase modified alternately by the erosional downcutting. Both depositional and depositional-erosional sequences vary in time and space. This is shown in the vicinity of the Baekdo Island where both phases are present (fig. 6.5a). This also suggests the instability of the moat. In the vicinity of the Samdo Island, the reflectors are continuous exhibiting only a depositional sequence (fig. 6.5b). Here, the mid-reflector also is laterally continuous filling the incised acoustic basement on both sides of the moat, which seems to have been stable for some time. The depressions in the southern tidal inlets show exclusively the depositional-erosional sequence with the discontinuous mid-reflector. The moats of both depositional and depositional-erosional origin appear to be similar to those of the large-scale, deep-sea moats on the Line Islands archipelagic apron reported by Normark and Spiess (1976).

The thick sediments contain certain gases encountered on the Uniboom records and are characterized by lack of acoustic penetration (fig. 2.18). These are similar to the acoustically turbid sediments in other shallow water environments (Reeburgh 1969; Schubel 1974). In Gamagyang Bay, the gas-charged zone thus occurs mainly in the central part comprising about 12 km^2 in area. Schubel (1974) and others attributed the acoustically turbid character to the scattering and attenuation of energy in the sediments which contain gas bubbles in their interstices. D'Olier (1979) also ascribed the "bright spot" reflections such as figure 6.7 to either the highly organic muds and gases or the thick accumulation of shells in the estuaries.

120 Marine Geology of Korean Seas

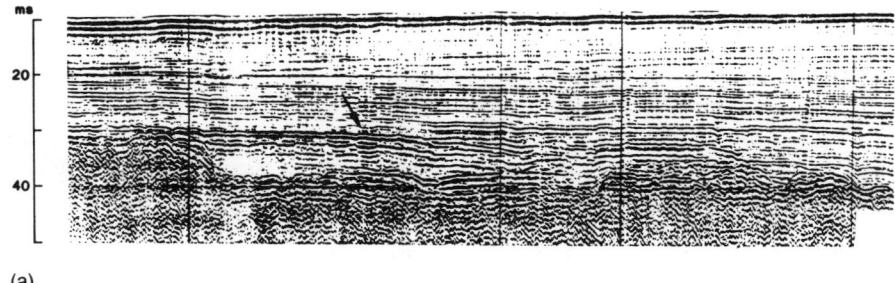

(a)

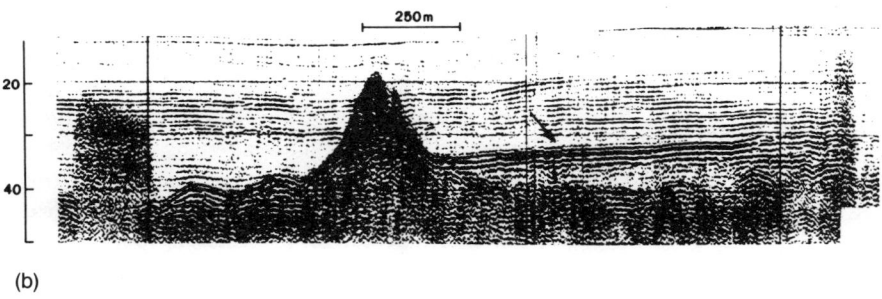

(b)

Figure 6.7. Uniboom profile across the eastern margin (west of Dolsan Island) of Gamagyang Bay showing mid-reflectors (arrows) at a depth of about 30 ms interval (for positions see fig. 6.1). Vertical scale: two-way traveltime in milliseconds; 10 milliseconds correspond to approximately 8 m sediment thickness. The mid-reflectors represent probably the erosional surface prior to the Holocene transgression about 4500 y B.P. The turbid layer (bright spot) on both ends of (b) represents the gas-charged sediments through which acoustic energy attenuates. After H.J. Kang and Chough (1982) by permission of Marine Geology, Elsevier Scientific Publishing Co.

The sediments in the gas-charged zone are also very fine, not different from the adjacent sea floor which consists of clayey silt. It is implied that the gases consist of CH_4, CO_2, and H_2S produced by the biochemical degradation of organic matter in the sediments.

According to the C-14 age determination using the shell fragments in two cores located on the northern and southern flanks of a depositional moat near Samdo Island, respectively, the sediments have accumulated at a rate of 134 cm/10^3 y (1645 ±200 y B.P. at a depth of 220 cm from the sea floor). Extrapolated at a constant rate, it yields an age of about 4500 y B.P. for the mid-reflector found approximately 6 m beneath the sea floor.

Sedimentary Structures

The sediments are bioturbated extensively and mottled by bottom-dwelling organisms (figs. 6.8a and 6.8b). Shell fragments as well as other organisms such as foraminifera, diatoms, radiolaria, spicules, and plant fragments are scattered throughout the core sections. The lack of primary sedimentary structures may be due to the fineness of the sediments. However, sediment inclusions of materials dissimilar to the matrix and the lenticular bedding, as described by Reineck and Wunderlich (1968), are occasionally found (fig. 6.8c).

Deposition of Fine Sediment

The sedimentary processes in the bay appear to be dominated by suspension transport of fine-grained sediments. The traction process seems to have played a minor role. On the average, the tidal currents during the mid-half of each cycle range from 25 to 35 cm/s in the central part of the bay (fig. 6.2). A maximum velocity of 72.4 cm/s was measured in the southern inlet during the ebb stage. The shear velocity (U_*) at the bed near 9 m exceeds 0.8 cm/s, which is large enough to move a particle of smaller than 6.5 ϕ in suspension. Most particles found in Gamagyang Bay which are finer than 6.5 ϕ in mean size will be transported in suspension during the major period of tidal currents. Deposition may occur only during the period approaching the full flood and ebb stages during which the shear velocity of tidal currents is substantially reduced probably below that of settling velocity.

The settling from suspension and the lack of traction probably cause the lack of primary sedimentary structures.

The progressive northward decrease of grain size (fig. 6.4) suggests that the influx of suspended sediments has occurred through the southern tidal inlets in which finer sediments travel farther northward and settle on the bottom probably due to settling and scouring lag effects. A lack of distinctive bed forms and of primary sedimentary structures may be caused alternatively by the fineness of sediments and the ultimate destruction by bottom-dwelling organisms. The lenticular bedding or coarsely interlayered bedding found occasionally in a number of cores is typical of the subtidal environment where mud deposition dominates. The beds are normally thin and seem to be formed by the occasional yield of sand or silt during the sporadic heavy storms, each lasting for a few days. The coarsely interlayered bedding may form when considerable amounts of sand or silt are available, whereas the flat isolated lenticular bedding may represent a meager supply of sand or silt.

The mechanics of moat formation may be shown by the Bernoulli equation. On approaching an island or a point, the tidal currents are deflected, causing an increase of current velocity. The pressure gradient, therefore, increases normal to the closer streamlines, precluding the deposition. The net deposition of sediments near the island or point probably occurs only when the current velocity reaches a minimal stage. The decrease of mean grain size toward the center of the moat may be due to the segregation of sediments by an eddy or a secondary helical flow. The complexity of depositional-erosional features shown in some Uniboom profiles suggests that the current direction has shifted at various stages of moat formation.

Deugryang Bay

Physiography

Deugryang Bay (590 km^2 in area) is also located on the southern coast, west of Gamagyang Bay (fig. 6.1). The bay is bounded by relatively high mountains and hills more than 400 m above sea level. Small-scale streams and creeks drain mainly into the western side of the bay forming a broad and shallow sandy shore. They also form sand spits at the mouth, some of which are transported northward along the shore. Some sediments in the north of Deugryang Island seem to have originated from discharge of sediments by these streams (J.H. Chang et al. 1980). The drainage system is limited on the east resulting in a rocky shore and gravelly coast. A tidal flat also is extensive locally, ranging up to about 3 km in width near the outlet of the bay.

The bay is also shallow (less than about 15 m) except for the tidal channel (or moat near the central island) (fig. 6.9). These scours or moats are similar to those described earlier in the Gamagyang Bay. The bay is characterized by semidiurnal tide flowing largely north and south. Maximum flood and ebb tidal currents attain a velocity of up to 2 knots during the spring tide (National Hydrographic Office of Korea 1980).

Sediments

The sediments are composed largely of poorly sorted muds (fig. 6.10) dominated by 6–7 ϕ size. In the scours and the tidal channels, however, they are slightly coarser (4–5 ϕ), and often the basement rocks are exposed on the sea floor. The average organic carbon in the bulk sediments is 0.37%. The sediments in the tidal channels are variable in grain size and usually contain granule-size materials which contain fragments of gneiss and volcanic rocks of local origin of the drainage area. Here, the sediments contain high $CaCO_3$ content (24.3–26.8%) composed largely of coarse-grained shell fragments, whereas that of organic carbon is decreased.

The less-than-2 μm clay minerals include kaolinite and chlorite (25–36%), illite (63–73%), and trace amounts of montmorillonite as well as sepiolite. The distribution of these minerals together with a slight increase of montmorillonite seaward is suggestive of significant landward transport of fine-grained materials.

Acoustic Characterisitics

The acoustic basement in Deugryang Bay forms a broad and high terrace below 15–28 m from sea level, except for some irregular relief in the central part of the bay (fig. 6.11). The high terrace lacks sediment on it. The uneven surface probably resulted from a differential erosion or also a fault trending both northeast- and northwest-ward.

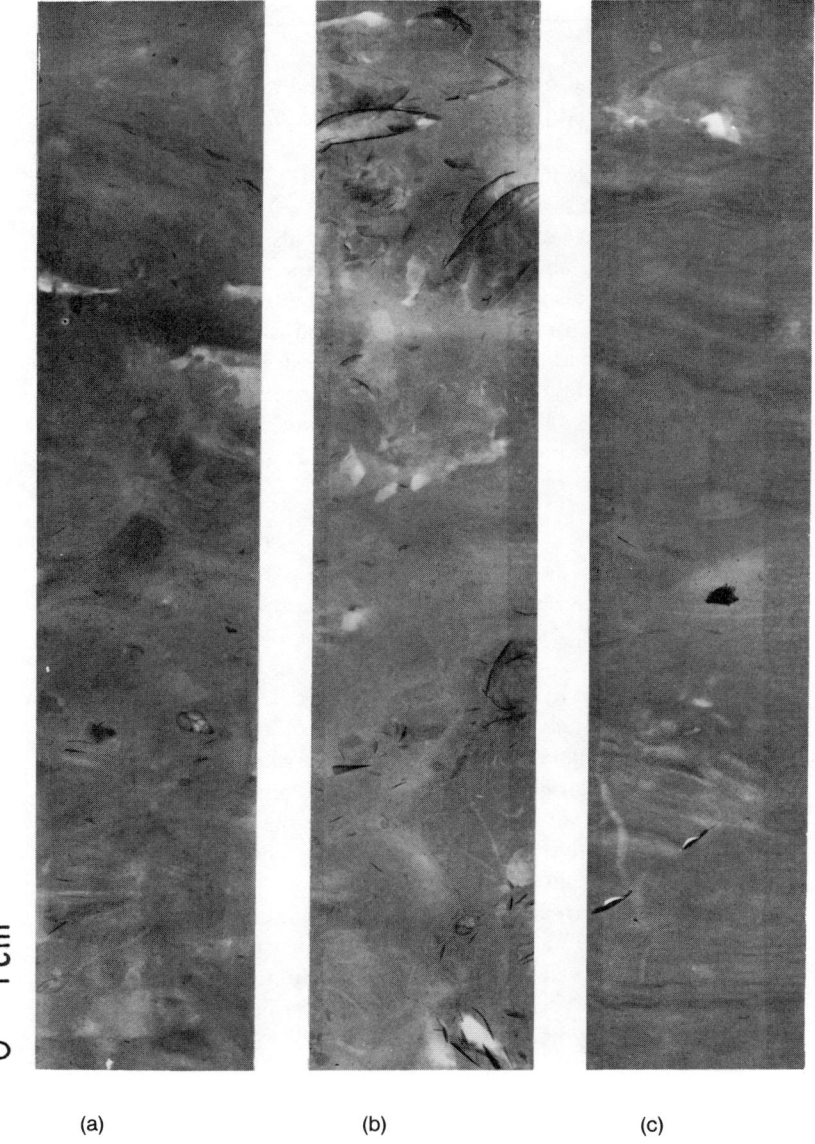

(a) (b) (c)

Figure 6.8. X-radiographs of sliced (1 cm thick) cores. a: core No. 4, core depth, 0–15 cm; b: core No. 6, core depth, 30–45 cm; c: core No. 9, core depth, 130–145 cm. Extensive bioturbation occurs in (a) and (b); large fragments are mostly broken shells. Irregular inclusions seen on (c) may result from unusual events such as storms that affect the shallow bay bottom.

Coastal Embayments (South Coast)   125

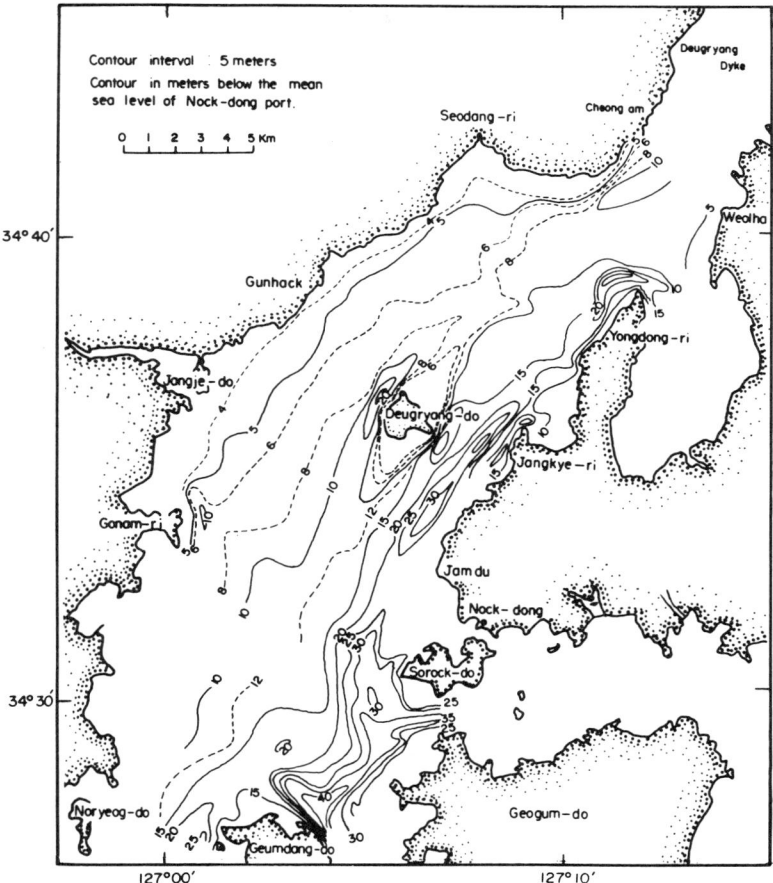

Figure 6.9. *Bathymetry of Deugryang Bay, corrected in mean sea level. For location of the bay see figure 6.1. After J.H. Chang et al. (1980) courtesy of the Korea Institute of Geosciences and Mineral Resources Report.*

126 Marine Geology of Korean Seas

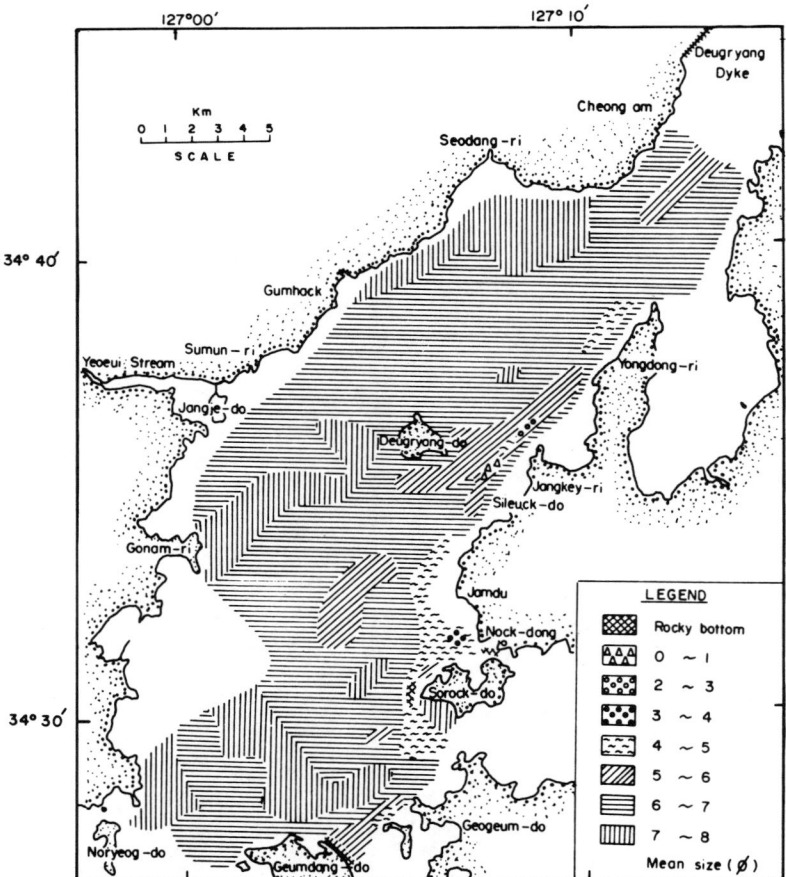

Figure 6.10. Grain-size distribution in Deugryang Bay. After J.H. Chang et al. (1980) courtesy of the Korea Institute of Geosciences and Mineral Resources Report.

Coastal Embayments (South Coast) 127

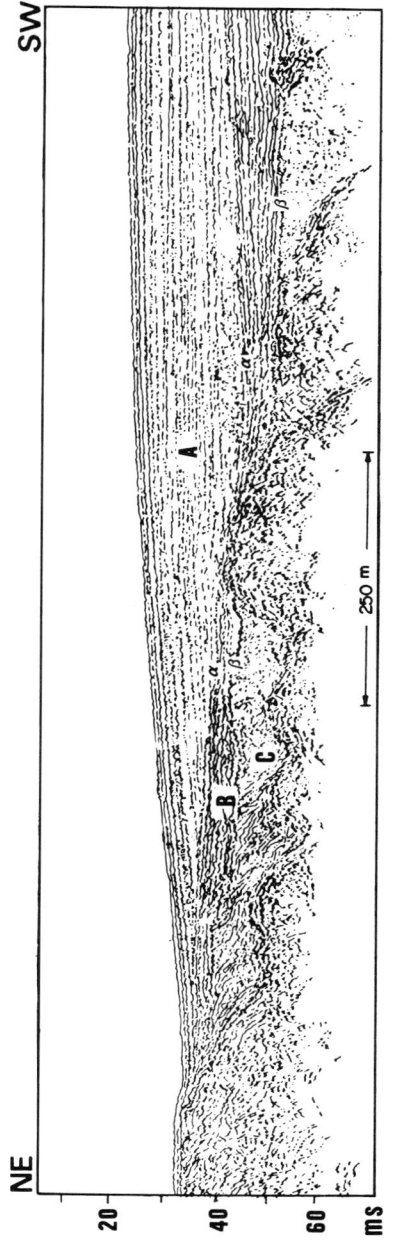

Figure 6.11. Uniboom profile across Deugryang Bay showing the mid-reflectors (α and β) and the sequences A, B, and C. Transgressive sequence A exhibits a coastal onlap. Vertical scale: two-way traveltime in milliseconds; 10 milliseconds correspond to approximately 8 m sediment thickness. Courtesy of the Korea Institute of Energy and Resources.

The overlying sediments are thicker generally in the central part than in the margin, which is coincident also with the low-relief, irregular basement area that trend both northeast- and northwest-ward. In the sediment column overlying the low-relief basement appears a mid-reflector that is certainly the same as in the Gamagyang Bay. Although the sediment layer below the mid-reflector was interpreted by J.H. Chang et al. (1980) as being nonmarine sediments deposited by mass wasting prior to the postglacial transgression, it is equally plausible that the layer is the erosional remnant of the pre-Wisconsin marine sediments that were deposited in the lower-lying area along the coast. The sediment layers on Uniboom profiles appear to be bedded, but they are usually homogeneous and in sediment cores they lack completely horizontal layers. The lack of sedimentary structures is most likely due to the deposition from suspension by tidal currents.

Late Quaternary History

The existence of a mid-reflector in Gamagyang Bay at the depth of about 20 to 25 m below the sea floor, representing an age older than about 4500 y B.P., suggests that the sedimentary sequence beneath it originated either from the subaerial mass wasting prior to the onset of recent transgression or from the erosion by strong tidal currents during the Holocene. On the other hand, the mid-reflector may signal the subaerial exposure of remnant marine sediments deposited prior to the late Wisconsinan time.

The same type of mid-reflector has also been formed in many embayments along the southern coast of the Korean Peninsula (Song and Cho 1978; J.H. Chang et al. 1980). Hahn and Kim (1977) reported a subaerial sand and gravel layer formed prior to the recent marine transgression at a depth of 16.5 m below the sea floor (about 13 cm above the basement) in a core retrieved in the northwestern part of the Jinhae Bay (fig. 6.1). Here, sediments were accumulated at a rate of about 113 cm/10^3 y (Hahn and Kim 1977). The same kind of material may cause the mid-reflector in Deugryang Bay (J.H. Chang et al. 1980). The mid-reflectors and overlying sedimentary sequence in areas such as Gamagyang and nearby Deugryang bays show evidence of onlap over the acoustic basement (fig. 6.11). This suggests marine transgression.

Further south, at the present water depth of about 40 to 60 m, the sediment sequence beneath the mid-reflector is thicker, and its top surface is eroded and truncated by the numerous erosional channels (fig. 6.12). These paleochannels were probably the extensions of the rivers and streams formed during Wisconsinan time when sea level was lower, eroding the pre-mid-reflector sequence. The pre-mid-reflector sequence thins progressively north-ward in the southern Korean coast suggesting that sea level stood near the bottom of most embayments during the pre-Wisconsinan interglacial period, barely depositing sediments in the depressed area.

In summary, the bays along the south coast seem to have been submerged during the pre-Wisconsinan interglacial period and subsequently exposed subaerially or lay above the depositional base level during the last glacial period. This is when the mid-reflector was formed on the surface of sedimentary sequence which filled preferentially the paleovalley and depressed basement prior to the last glacial period. During the postglacial transgression (Y.A. Park 1969; Emery et al. 1971), the bays were submerged again, but little or no deposition occurred until about 4500 y B.P. prior to which the sea level rose rapidly.

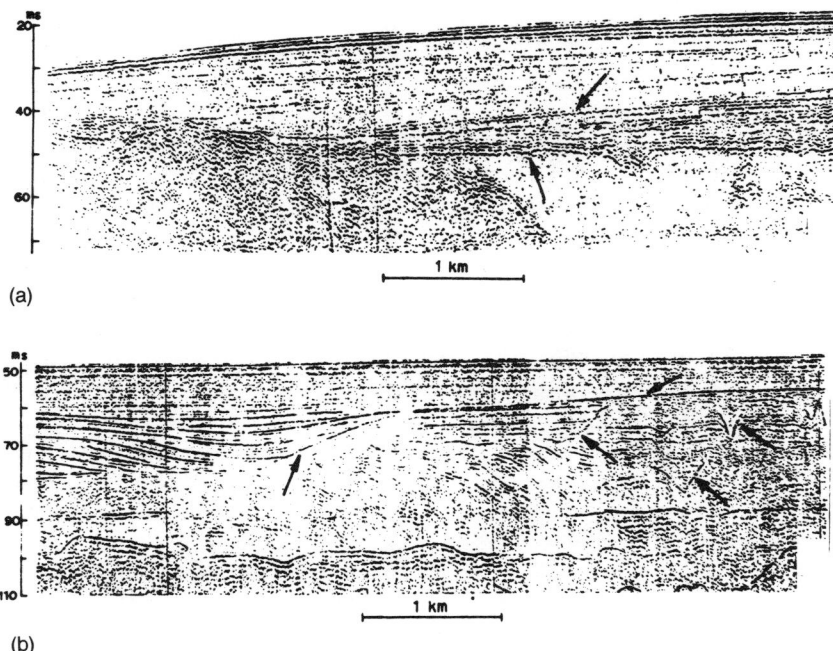

Figure 6.12. Uniboom profile of the South Sea between the Geoje and Namhae islands showing the truncated mid-reflectors and erosional paleochannels (arrows). Vertical scale: two-way traveltime in milliseconds; 10 milliseconds correspond to approximately 8 m sediment thickness. Courtesy of the Korea Institute of Energy and Resources.

7 Geological History

General Statement

Geological information obtained in the seas around the Korean Peninsula is important for interpreting the geological history of the Korean Peninsula on land and under the sea. The origin and development of the Korean Seas are related to the plate motions and their tectonic regime (Koo et al. 1980a). Sedimentation and deformation in the intracratonal Yellow Sea are thus different from those in the East Sea, the latter situated near the plate boundary. However, global-scale reconstructions of plate motions often deviate from the local phenomena and thus are incoherent. This becomes further complicated with pre-Mesozoic plate motions. B.K. Park (1977) suggested that the Korean Peninsula was the margin of Pangaea prior to the Mesozoic.

The tectonic force to rift the pre-Mesozoic basins in the peninsula may result from various sources depending on the type and location relative to the plate position. Although this argument can best be resolved by studying the outcrops of land, the geological information obtained under the sea provide data on basin configuration, tectonic regime, and sedimentation processes. These and other problems that are pertinent to the origin and development of the seas around the Korean Peninsula are summarized next in a geochronological order.

Precambrian

The Yellow Sea is underlain most likely by the stable craton of the Sino-Korean Platform (Precambrian). The crust is approximately 35 km thick and typically of continental crust (5.8 km/s) under the Korean Peninsula (K. Lee 1979). Although sedimentation as well as plutonism and metamorphism took place during the early to middle Precambrian period (Reedman and Um 1975), many basins in this period probably were intracratonic. However, the origin and nature of sedimentation of original paragneiss as well as schist are speculative at present.

The stable craton was rifted undergoing progressive subsidence. This rift zone trends northeast–southwest in the middle part of the southern Korean Peninsula and is now known as the

Ogcheon Fold Belt, separating the Gyeonggi Massif from the Yeongnam Massif. The initiation of the Ogcheon Basin was as early as late Proterozoic (O.J. Kim 1970; J.H. Lee 1972; Reedman and Um 1975) or post-Cambrian (K.W. Kim and Lee 1965; Son 1970), and was compared either to the classical geosyncline (D.S. Lee 1980) or an island arc (B.K. Park and So 1972).

The Ogcheon Basin, more or less symmetrical in cross section, deposited sediments in shallow marine facies on either side. It underwent progressive downwarping, resulting in debris flow deposits associated with occasional limestone sequences toward the shelf and mudstone offshore. This symmetry and the characteristics of sedimentary facies suggest an aulacogen origin (Chough 1981b; Chough et al. 1981a). The basin was partly isolated by an arch in the rift resulting in an anoxic basin. This is indicated by the extensive occurrence of euxinic shale and lack of marine organisms. On land, the debris flow deposits were confined to the northern part of the basin separated by an arch near Daejeon. In modern ocean basins, debris flow processes (Hampton 1972) are prevalent on the continental slope and base-of-slope environments (Embley 1976; Prior and Coleman 1977; Nardin et al. 1979) where the submarine slope failure occurs frequently due to abnormally high pore pressure or sudden shocks.

The fact that mafic rocks (So and Kim 1975; H.Y. Lee et al. 1980; D.S. Lee 1980) occur in association mostly with the rocks of shallow water origin in the rim of the basin is suggestive of their emplacement during the early phase of basin rift and downwarping in a shallow marine or near isolated islands. The mafic rocks in the Ogcheon Basin are not associated with the sequence of typical ophiolite suite such as pillow breccia, pelagites, and flysch deposits. Although Gnibidenko (1979) suggested that the Ogcheon Fold Belt extends to the continental margin of Siberia across the Japan Basin the evidence for this is minimal.

The Japanese Island was situated closer to the Asian continent, and drifted southward most probably in Miocene time.

Paleozoic

Possible epeirogenic crustal movements took place during the early Cambrian to Triassic (Son 1969) in the cratonic interior, the sea invading the northeastern part of the Ogcheon Basin. The advance of the sea resulted in extensive deposition of carbonates whose extension into the East Sea margin has yet to be outlined. During late Paleozoic time, deposition of sediments in shallow marine and nonmarine environments prevailed in the area accompanied by an extensive epeirogenic movement. The sedimentation in the Ogcheon Basin seems to have continued into the Paleozoic time until deformed in the late Mesozoic Daebo Orogeny. The Yellow Sea probably remained the same in the Precambrian time.

Mesozoic

The configuration of the Yellow Sea and that of the Ogcheon Basin was much the same in the Precambrian as the Paleozoic. A series of orogenic movement occurred in the Triassic (Songrim Disturbance) followed by Jurassic (Daebo) Orogeny. The Yellow Sea region was probably uplifted after this series of deformation resulting in the nondeposition or postorogenic inland molasse sedimentation. The orogenies resulted in a series of northeast–southwest trending ridges that blocked sediment transport to the southeast. One of the ridges is the extension of the Ogcheon Fold Belt and Yeongnam Massif on land, and the other in the southern Yellow Sea (or northern East China Sea) south of Jeju-Yangtze line. The latter blocked the sediments west of Taiwan-Sinzi Zone, depositing thick Tertiary sediments.

In the East Sea, the protosea was initiated probably in Cretaceous time with the subduction of the Kula-Pacific Ridge under the Asian continental plate (Uyeda and Miyashiro 1974; Hilde et al. 1976). According to Uyeda (1979) and Uyeda and Kanamori (1979), the sea was rifted by small-scale convection associated with subduction below the back arc. The opening of the sea was probably in two discrete steps: first rifting of the Japan Basin followed later by the Yamato and Ulleung basins. The Precambrian gneiss in the Yamato Ridge is the remnant of continental crust replaced in the basin floor by the oceanic crust formed by the back arc spreading.

The spreading was from northwest to southeast as indicated by the paleomagnetic data on the northwestern margin of

Honshu and the northwest–southeast trending block faulting in the western margin of the Ulleung Basin. The faults were formed in late Miocene time and barred sediment transport further east during the late Neogene. It was later replaced by extensive sedimentation in the form of mass flows resulting in the rapid filling of the deep basins by turbidites. On the Korean shelf, sediment accumulation was minimal since this time when the rifted block was uplifted progressively.

The subduction of the Kula-Pacific Ridge probably triggered tectonic instability resulting in a series of orogenies in the mid-Jurassic time associated with extensive magmatism. Cretaceous volcanics erupted along the southern coast of Korean Peninsula associated with the intrusion of granite. The Eurasian Plate moved to the southeast.

Cenozoic

During the Cenozoic, the Eurasian Plate motion was to the southeast (Terman 1977). While the Pacific Plate subducts along the trenches under the Eurasian Plate, the Eurasian Plate was bound by a number of regions which behaved separately (fig. 7.1). The Marine East Asia region actively fed the subduction zones of the Pacific Plate. The North-Central East Asia Plate on which the Korean Peninsula and Seas occur is characterized by many graben, alkalic volcanism, and shallow seismicity. This is indicative of crustal extension during the Cenozoic (Terman 1977). Associated with this extension is the opening of back-arc basins such as Ulleung Basin and Okinawa Trough as peripheral fragments of continental crust migrate away from Asia.

Continental sedimentation was prevalent in the Yellow Sea filling the postorogenic depressions. It was not until late Pliocene that the sea advanced into the Yellow Sea region, depositing unconformably on the deformed nonmarine sediments that were extensive on the west, barely filling the Heugsan Platform.

In the East Sea, the sea advanced into the rifted area as early as late Cretaceous or late Oligocene. During the initial phase of rift it consists of paralic sediments including volcanics, followed by progressively deeper marine facies. Turbidite and the associated mass-flow sedimentation commenced in the Ulleung Basin

Geological History 135

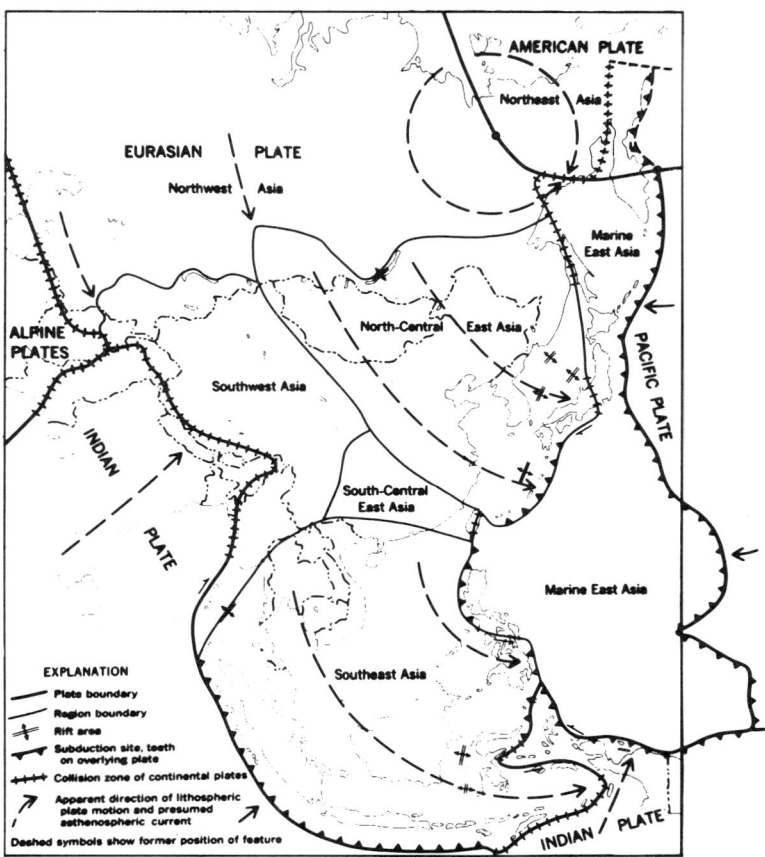

Figure 7.1. Cenozoic tectonics of East Asia. After Terman (1977) by permission of the American Geophysical Union.

Basin and in the Korea Strait. The rifting or back-arc opening ceased by about late Neogene time; and the Tertiary sequence here was folded, block faulted, and uplifted since this time in the presently continental margin area.

Late Quaternary

Since about Pliocene time, the major part of the Korean Seas was below sea level; the Yellow Sea as an epicontinental sea in a stable cratonic area and the East Sea as a deep marginal sea whose spreading ceased sometime in the late Neogene. The East Sea margin of the Korea Peninsula continues to emerge and most sediments are delivered to the deep basins (Ulleung Basin). The turbidity current activity in the Ulleung Basin was enhanced during the glacial period when the sea level was lower, generated via mass flows of line source. The basin was isolated from the Pacific by the shallow sills during this time. An inflow of Pacific water mass began with the postglacial transgression and cooled in the northern part of the sea to form bottom currents of highly oxygenated water.

On the western coast significant amounts of sediments also were accumulated during the Quaternary period. During the postglacial, sea-level rise, the coastline and tidal marsh were submerged (Y.A. Park 1969). The existence of raised spits exposed on the coast 2-3 m above the present highest sea level (Youn et al. 1977) suggest, however, that sea level fluctuated since the postglacial transgression. On the southeast coast, the raised beaches (or terraces) are indicative of uplift, at a rate of 1.1 to 1.4 mm/y during late Pleistocene (S.W. Kim 1973).

At the present water depth of less than about 100 m in the Yellow Sea, the sediment sequence beneath the mid-reflector α is thicker toward offshore, whose top surface was eroded and truncated by numerous erosional channels. These paleochannels were the extensions of the rivers and streams formed during Wisconsin time when sea level was lower.

Sedimentation in the Yellow Sea during the postglacial transgression was dominated by the discharge of both the Yangtze and the Hwangho rivers. Deposition in the southeastern Yellow

Sea was controlled by prevailing currents and tides. Erosion occurred largely in the areas adjacent to the topographic high or islands affected by strong tidal currents. Here, pre-Wisconsinan sediments (layers B or C) were exposed to the sea floor.

The existence of a mid-reflector (older than about 4500 y B.P.) in the Gamagyang Bay at a depth of about 20 to 25 m below the sea floor may signal the subaerial exposure of remnant marine sediments deposited prior to late Wisconsin time. The same type of mid-reflector, also found in many embayments along the southern coast, composed of subaerial sand and gravel layer, shows evidence of onlap suggestive of marine transgression. Sea level stood near the bottom of most embayments during the pre-Wisconsin interglacial period, barely depositing sediments in the depressed area. Deposition began probably when the sea level rose enough to cover the bay with a considerable water depth placing the bay below the depositional base level.

References

Aoki, S., and K. Oinuma, 1973, Clay minerals in the sediments of the continental shelf, off San-in, the Japan Sea: The Jour. of Assoc. for Geologists Collaboration in Japan, v. 27, p. 35-39.

Aoki, S., K. Oinuma, and T. Sudo, 1974, The distribution of clay minerals in the recent sediments of the Japan Sea: Deep-Sea Res., v. 21, p. 299-310.

Arai, F., T. Oba, H. Kitazato, Y. Horibe, and H. Machida, 1981, Late Quaternary tephrochronology and paleo-oceanography of the sediments of the Japan Sea: The Quat. Res., v. 20, p. 209-230.

Asano, K., 1957, The foraminifera from the adjacent seas of Japan, collected by the S.S. Soyo Maru, 1922-1930: Sci. Rept. Tohoku Univ., ser. 2 (Geol.), v. 28, p. 1-52.

Aubert, H., and M. Pinta, 1977, Trace Elements in Soils: Amsterdam, Elsevier Scientific Publ. Co., 395 p.

Bahk, K.S., 1982, Provenance of turbidites in the Ulleung back-arc basin, East Sea: Unpubl. Master's thesis, Seoul National Univ., Seoul, Korea, 77 p.

Bahk, K.S., and S.K. Chough, 1983, Provenance of turbidites in the Ulleung (Tsushima) back-arc basin, East Sea (Sea of Japan): Jour. Sed. Pet., v. 53.

Beresnev, A.F., and V.M. Kovylin, 1970, Basement relief and thickness of bottom sediments in the central part of the Sea of Japan: Okeanologiya, v. 10, no. 1, p. 86-88.

Bersenev, I.I., 1971, The origin of the Japan Sea Basin, in Island arc and marginal sea: Tokyo, Japan, Tokai Univ. Press, p. 31-37.

Blatt, H., G. Middleton and R. Murray, 1980, Origin of Sedimentary Rocks: 2nd ed., Englewood Cliffs, New Jersey, Prentice-Hall, 782 p.

Bosum, W., E.G. Kind, and J.H. Koo, 1971, Aeromagnetic survey of offshore areas adjoining the Korea Peninsula: U.N. ECAFE, CCOP, Tech. Bull., v. 4, p. 1-21.

Chang, J.H., S.W. Kim, and Y.A. Park, 1978, Sedimentologic natures of the bottom sediments between Mokpo and Cheongsan Island off the southern coast of Korea: Jour. Geol. Soc. Korea, v. 14, no. 2, p. 17-23.

Chang, J.H., C.W. Lee, K.S. Park, W.S. Kim, and W.C. Shin, 1980, Geophysical and geological study for Quaternary mineral resources in Deugryang Bay, southern coast, Korea: Rept. Geosci. and Min. Resour., Korea Inst. Geosci. and Min. Resour. (KIGAM), v. 9, p. 35-63.

Chang, K.H., 1975, Unconformity-bounded stratigraphic units: Geol. Soc. Am. Bull. v. 86, p. 1544-1552.

Chang, K.H. 1977, Late Mesozoic stratigraphy, sedimentation and tectonics of southeastern Korea: Jour. Geol. Soc. Korea, v. 13, p. 76-90.

Chang, K.H. and H.M. Kim, 1968, Cretaceous paleocurrents and sedimentation in northwestern part of the Kyeongsang Basin, southern Korea: Jour. Geol. Soc. Korea, v. 2, p. 21-38.

Chang, S.K. and B.K. Kim, 1976, Foraminifera in the Maro Sea off Jindo Island, Korea: Jour. Geol. Soc. Korea; v. 12, no. 1, p. 31-54.

Chen, P.Y., 1978, Minerals in bottom sediments of the South China Sea: Geol. Soc. Am. Bull., v. 89, p. 211-222.

Cheong, C.H., 1982, Some suggestions on the stratigraphic units in Korea, in Mem. J.H. Kim's 70th Birthday: Korea Mining Productivity Center, Seoul, Korea, p. 133-152.

Chinhae Machine Depot, 1979, Korea East Sea Coast, Bottom Distribution Map of Recent Sediments: scale 1:500,000.

Cho, C.J. and H.I. Choi, 1970, Report on geology of islands off the south coast of Korea: Rept. Mar. Geol. and Geophys., Geol. Surv. Korea, no. 1, p. 41-78.

Cho, K.J., 1979, Geophysical study on the geological structrue of the South Sea of Korea: Unpubl. PhD thesis, Hanyang Univ., Seoul, Korea, 107 p.

Choe, K.L., 1981, Faunal analysis of recent ostracodes and benthonic foraminifers in the bottom sediments from Jindo Island to Jeju Island (Jeju Strait), Korea: Unpubl. Master's thesis, Korea Univ., Seoul, Korea, 100 p.

Choi, H.I., 1981, Depositional environments of the Sindong Group in the southwestern part of the Gyeongsang Basin: Unpubl. PhD thesis, Seoul National Univ., Seoul, Korea, 144 p.

Choi, H.I. and S.D. Hahn, 1975, Marine geological study of the Keum river estuary: Rept. on Geol. and Mineral, Geol. and Mineral Inst. Korea, II, no. 10, p. 79-105.

Choi, J.H., 1981, Recent clay minerals in the Kunsan Estuary and the adjacent continental shelf: Unpubl. Master's thesis, Seoul National Univ., Seoul, Korea, 52 p.

Chough, S.K., 1981a, Dispersal of the southeastern Yellow Sea sediments: a steady-state model: Proc., CCOP, 18th session, Seoul, Korea, p. 130-142.

Chough, S.K., 1981b, Submarine debris flow deposits in the Ogcheon Basin, Korean Peninsula: U.N. ESCAP, CCOP, Tech. Bull., v. 14, p. 17-29.

Chough, S.K. 1982, Turbidites in the Ulleung (Tsushima) back-arc basin, East Sea (Sea of Japan), in M. Hoshino and T. Shibasaki, eds., Geology of Japan Sea: Tokyo, Japan, Tokai Univ. Press, p. 365-376.

Chough, S.K., 1983a, Further evidence of fine-grained sediment dispersal in the southeastern Yellow Sea: Sedimentary Geology.

Chough, S.K. 1983b, Fine-grained trubidites and associated mass flow deposits in the Ulleung (Tsushima) Basin, East Sea (Sea of Japan), in D.A.V. Stow and D.J.W. Piper, eds., Fine-grained sediments, deep water processes and environment: Geol. Soc. of London Special Publication

Chough, S.K. and K.S. Jeong, in prep., Zoned facies of mass flow deposits in the Ulleung Back-Arc Basin.

Chough, S.K., H.J. Kang, K.S. Bahk, J.H. Kim, and J.K. Park, 1981a, Submarine debris flow deposits in the Ogcheon Basin (? Late Proterozoic-Cretaceous), Korean Peninsula: preliminary results: Proc. CCOP, 18th session, Seoul, Korea.

Chough, S.K. and D.C. Kim, 1981, Dispersal of fine-grained sediments in the southeastern Yellow Sea: A steady-state model: Jour. Sed. Petrol. v. 51, p. 721-728.

Chough, S.K., K. Kim, and H.J. Kang, 1982, Deposition of fine-grained sediments in tide-dominated embayment, Gamagyang Bay, southern coast of Korea: Korea Ocean Res. and Devel. Inst. Bull. BSPE 00028-51-3. p. 37-74.

Chough, S.K., K. Tamaki, K.S. Bahk, E. Inoue, and M. Yuasa, 1981b, Heavy minerals from the Oki Spur, Japan Sea: Bull. Geol. Surv. Japan, v. 32, p. 487-501.

Cook, H.E., M.E. Field, and J.V. Gardner, 1982, Continental slopes, in P.A. Scholle and D. Spearing, eds., Sandstone depositional environments: AAPG Mem. 31, p. 329-364.

Dickinson, W.R. and R. Valloni, 1980, Plate setting and provenance of sands in modern ocean basins: Geology, v. 8, p. 82-86.

D'Olier, B., 1979, Side-scan sonar and reflection seismic profiling, in K.R. Dyer, ed., Estuarine hydrography and sedimentation: Cambridge, England, Cambridge Univ. Press, p. 57-86.

Embley, R.W., 1976, New evidence for occurrence of debris flow deposits in the deep sea: Geology, v. 4, p. 371-374.

Emery, K.O., Y. Hayashi, T.W.C. Hilde, K. Kobayashi, J.H. Koo, C.Y. Meng, H. Niino, J.H. Osterhagen, L.M. Reynolds, J.M. Wageman, C.S. Wang, and S.J. Yang, 1969, Geological structure and some water characteristics of the East China Sea and the Yellow Sea: U.N. ECAFE CCOP, Tech. Bull., v. 2., p. 3-43.

Emery, K.O. and H. Niino, 1967, Stratigraphy and petroleum prospects of Korea Strait and the East China Sea: Rept. Geophys. Explor., Geol. Surv. Korea, v. 1, p. 249-263.

Emery, K.O., H. Niino, and B. Sullivan, 1971, Post-Pleistocene levels of the East China Sea, in K.K. Turekian, ed., The Late Cenozoic Glacial Ages: New Haven, Ct., Yale Univ. Press, p. 381-390.

Folk, R.L., 1968, Petrology of sedimentary rocks: Austin, Texas, Hemphill's, 170 p.

Frazier, S.B., S.O. Choi, B.K. Kim and D. Schwartz, 1976, Marine petroleum exploration of Huksan platform, Korea, in M.T. Halbouty, J.C. Maher, and H.M. Lian, eds., Circum-Pacific energy and mineral resources: AAPG Mem. 25, p. 268-275.

Fullagar, P.D. and B.K. Park, 1975, R_b-R_r study of granite and gneiss from Seoul, South Korea: Geol. Soc. Am. Bull., v. 86, p. 1579-1580.

Gnibidenko, H., 1979, The tectonics of the Japan Sea: Mar. Geol., v. 32, p. 71-87.

Gorai, M., 1968, Some geological problems in the development of Japan and the neighboring island arcs, in C.L. Drake and L.L. Knopoff, eds., The crust and upper mantle of the Pacific Area: AGU Monogr. no. 12, p. 481-485.

Gorai, M. 1982, Reconsideration on the origin of oceanic crust of deep sea basins in the Sea of Japan, in M. Hoshino and T. Shibasaki, eds., Geology of Japan Sea: Tokyo, Japan, Tokai Univ. Press, p. 33-36.

Gradusov, B.P., 1974, A tentative study of clay mineral distribution in soils of the world: Geoderma, v. 12, p. 49-55.

Hahn, S.D., 1979, Bottom topography of southern seas of Korea: Korea Ocean Res. and Devel. Inst., scale 1: 1,500,000.

Hahn, S.D. and E.H. Kim, 1977, Coastal and marine geology of northwestern part of Jinhae Bay: Rept. Geosci. and Min. Resour., Korea Res. Inst. Geosci. and Min. Resour. (KIGAM), v. 1, p. 203-247.

Hahn, S.D., H.J. Lie, M.S. Suk, P.S. Park, H.K. Jun, and S.C. Hwang, 1978a, Horizontal temperature distributions of Korean waters (1961-1975), in S.D. Hahn, ed., Oceanographic atlas of Korean waters: Korea Ocean Res. and Devel. Inst., v. 1, p. 25-118.

Hahn, S.D., M.S. Suk, and P.S. Park, 1978b, Vertical temperature distributions and their variabilities of Korean waters (1961-1975), in S.D. Hahn, ed., Oceanographic atlas of Korean waters: Korea Ocean Res. and Devel. Inst., v. 1, p. 177-190.

Hampton, M.A., 1972, The role of subaqueous debris flow in generating turbidity currents: Jour. Sed. Pet. v. 42, p. 775-793.

Han, S.J. 1979, Clay minerals in recent sediments of the Korea Strait: Korea Ocean Res. and Devel. Inst. Bull., v. 1, no. 1, p. 23-37.

Hasegawa, Y., 1970, Diatom flora of gravity core samples from the Yamato Rise in the central part of the Japan Sea: Jour. Geol. Soc. Japan, v. 76, p. 347-354.

Hilde, T.W.C., S. Uyeda, and L. Kroenke, 1976, Evolution of the western Pacific and its margin: U.N. ECAPE, CCOP, Tech. Bull., v. 10, p. 1-19.

Hilde, T.W.C. and J.M. Wageman, 1973, Structure and origin of the Japan sea, in P.J. Coleman, ed., The Western Pacific: New York, Crane, Russak and Co.

Inc. and Univ. Western Australia Press, p. 413-434.
Honza, E., 1979a, Sediments, structure and origin of Japan Sea: concluding remarks, in E. Honza, ed., Geological investigation of the Japan Sea: Geol. Surv. Japan, Cruise Rept. no. 13, p. 89-93.
Honza, E., 1979b, Sediments, structure and spreading of Japan Sea: Nihonkai, no. 10, p. 23-45.
Honza, E., H. Kagami, and N. Nasu, 1977, Neogene geological history of the Tohoku Island System: Jour. Ocean. Soc. Japan, v. 33, p. 297-310.
Honza, E., K. Tamaki, M. Yuasa, and F. Murakami, 1979, Geological map of the southern Japan Sea and Tsushima Strait, 1:1,000,000: Geol. Surv. Japan.
Honza, E., M. Yuasa, and K. Ishibashi, 1978, Cored material, in E. Honza, ed., Geological investigations in the northern margin of the Okinawa Trough and the western margin of the Japan Sea: Geol. Surv. Japan, Cruise Rept. no. 10, p. 39-42.
Hoshino, M. and H. Homma, 1966, Geology of submarine banks in the Japan Sea: Chikyu Kagaku (Earth Sciences), no. 82, p. 10-16.
Huntec Ltd., 1968, Report of the offshore geophysical survey in the Pohang area, R.O.K.: U.N. ECAFE, CCOP, Tech. Bull., v. 1, p. 1-12.
Hurley, P.M., H.W. Fairbairn, W.H. Pinson, and J.H. Lee, 1973, Middle Precambrian and older apparent age values in basement gneisses of South Korea, and relations with southwest Japan: Geol. Soc. Am. Bull., v. 84, p. 2299-2304.
Ichikura, M. and H. Ujiié, 1976, Lithology and planktonic foraminifera of the Sea of Japan piston cores: Bull. Nat. Sci. Mus. Ser. (Geol.), v. 2, p. 151-178.
Ingle, J.C. Jr., 1975, Summary of late Paleogene-Neogene insular stratigraphy, paleobathymetry, and correlations, Philippine Sea and Sea of Japan region, in D.E. Karig et al., eds., Initial Reports of the Deep-Sea Drilling Project, U.S. Govt. Print. Off., v. 31, p. 837-855.
Isezaki, N., 1975, Possible spreading centers in the Japan Sea: Mar. Geophys. Res., v. 2, p. 265-277.
Isezaki, N. and S. Uyeda, 1973, Geomagnetic anomaly pattern of the Japan Sea: Mar. Geophys. Res., v. 2, p. 51-59.
Ishibashi, K. and E. Honza, 1978, Bathymetric survey, in E. Honza, ed., Geological investigations in the northern margin of the Okinawa Trough and the western margin of the Japan Sea: Geol. Surv. Japan, Cruise Rept., no. 10, p. 12-14.
Ishiwada, Y. and K. Ogawa, 1976, Petroleum geology of offshore areas around the Japanese Islands: U.N. ECAFE, CCOP, Bull., v. 10, p. 23-34.
Jacobi, R.D., 1976, Sediment slides on the northwestern continental margin of Africa: Mar. Geol., v. 22, p. 157-173.
Jeong, K.S., 1983, Mass flow deposits in the Ulleung Basin, East Sea: Unpubl. Master's thesis, Seoul National Univ. Seoul, Korea, 122 p.
Jin, M.S., 1981, Petrology and geochemistry of the Cretaceous granitic rocks in southern Korea: Unpubl. PhD thesis, Seoul National Univ., Seoul, Korea, 145 p.
Joshima, M., 1978, Gravity measurements, in E. Honza, ed., Geological investigations in the northern margin of the Okinawa Trough and the western margin of the Japan Sea: Geol. Surv. Japan, Cruise Rept., no. 10, p. 21-36.
Kang, H.J., 1981, Late Quaternary sedimentary processes in the Gamagyang Bay, southern coast of Korea: Unpubl. Master's thesis, Seoul National Univ., Seoul, Korea, 102 p.

Kang, H.J. and S.K. Chough, 1982, Gamagyang Bay, southern coast of Korea: sedimentation on a tide-dominated rocky embayment: Mar. Geol., v. 48, p. 197-214.

Kang, P.C. and K.H. Chi, 1980, Study on the geological structure of Ogcheon System using remotely-sensed data: Rept. Geosci. and Min. Resour., Korea Res. Inst. Geosci. and Min. Resour. (KIGAM), v. 8, p. 21-48.

Karig, D.E., 1971, Origin and development of marginal basins in the western Pacific: Jour. Geophy. Res., v. 76, p. 2542-2561.

Karig, D.E., J.C. Ingle, Jr., et al., 1975, Initial Reports of the Deep Sea Drilling Project: U.S. Govt. Printing Office, v. 31, 927 p.

Kaseno, Y., 1971, Cenozoic history of the Japan Sea coast region of Japan, with reference to the development of the Japan Sea, in Island Arc and Marginal Sea: Tokyo, Japan, Tokai Univ. Press, p. 1-4.

Kaseno, Y., 1972, Geological features of the Japan Sea floor: a review of recent studies: Pacific Geol., v. 4, p. 91-111.

Kato, M., 1978, Age assignment of the dredge and piston core samples, in E. Honza, ed., Geological investigations in the northern margin of the Okinawa Trough and the western margin of the Japan Sea: Geol. Surv. Japan, Cruise Rept., no. 10, p. 59-62.

Kim, B.K., 1965, The stratigraphic and paleontologic studies on the Tertiary (Miocene) of the Pohang area, Korea: Seoul National Univ., Jour. Sci. Tech., v. 15, p. 32-121.

Kim, B.K. and J.H. Han, 1971, Foraminifera in the bottom sediments off the southeastern coast of Korea: Jour. Geol. Soc. Korea, v. 7, no. 1, p. 11-36.

Kim, B.K. and J.H. Han, 1972, A foraminiferal study of the bottom sediments off the southeastern coast of Korea: U.N. ECAFE, CCOP, Tech. Bull., v. 6, p. 13-29.

Kim, B.K., S.W. Kim, and J.J. Kim, 1970, Foraminifera in the bottom sediments off the southwestern coast of Korea: U.N. ECAFE, CCOP, Tech. Bull., v. 3, p. 147-163.

Kim, C.G., S.W. Kim, M.H. Lee, and H.J. Lee, 1974, Study on the bottom sediments in the middle part of the Yellow Sea: Rept. Mar. Geol. Geophys., Geol. Min. Inst. Korea, v. 2, p. 95-118.

Kim, C.G., S.W. Kim, M.Y. Yang, and C.H. Chang, 1975, Study on the bottom sediments in the middle part of the Yellow Sea: Rept. Mar. Geol. Geophy., Geol. Min. Inst. Korea, v. 3, p. 1-54.

Kim, C.M. and W.Y. Lee, 1974a, Report on marine geophysical prospecting of Hamanri-Kunsan, west coast, Korea: Rept. of Geol. and Min. Explor., Geol. Surv. Korea, v. 2, p. 1, p. 155-182.

Kim, C.M. and W.Y. Lee, 1974b, Geophysical survey of the West Sea between 36-37°N, and east of 125°30'E. Rept. on Geol. and Min. Studies, Geol. Surv. Korea, no. 2, p. 155-160.

Kim, C.M. and W.C. Shin, 1981, Isopach map of surficial sediments: Submar. Geol. Map of Korean Cont. Shelf, Ser. II, Korea Inst. of Energy and Resour. (KIER), Map II-4.

Kim, C.M., W.C. Shin, W.S. Kim, and K.J. Cho, 1981, Marine geophysical survey off middle western coast of Korea between Gunsan and Buan: Rept. on Geosci. and Min. Resour., KIER, v. 11, p. 215-226.

Kim, C.M., J.J. Yang, and Y.S. Kang, 1971, Beach sediments of southeastern coast of Korea: Rept. Mar. Geol. and Geophys., Geol. Surv. Korea, no. 2, p. 89-110.

Kim, C.S., 1976, Petroleum potential of Korean offshore, in M.T. Halbouty, J.C. Maher, and H.M. Lian, eds., Circum-Pacific energy and mineral resources: AAPG Mem. 25, p. 261-267.

Kim, C.S., 1981, Submarine geology of continental margin of the East Sea: Unpubl. PhD thesis, Seoul National Univ., Seoul, Korea, 81 p.

Kim, C.S. and C.S. Kang, 1973, The geological study of shoreline and offshore in Youngil Bay and its vicinity: Rept. Geol. and Mineral Explor., Geol. Surv. Korea, v. 1, pt. 1, p. 97-111.

Kim, C.S., S.W. Kim, J.H. Chang, C.W. Lee, G.H. Min, C.M. Kim, W.S. Kim, and W.C. Shin, 1982, Geological and geophysical survey on the continental shelf off southwestern Korea: Korea Inst. Energy and Res. Bull., no. 31, 48 p.

Kim, C.S., J.H. Koo, C.M. Kim, M.H. Chang, and Y.K. Kim, 1972, Reconnaissance geophysical survey of southwest part of Yellow Sea: Geol. Surv. Korea Bull., v. 14, p. 620-658.

Kim, C.S., J.H. Koo, and S.J. Yang, 1969, Report on geophysical prospecting in the Yellow Sea and the East China Sea: Rept. of Geophys. Explor., Geol. Surv. Korea, v. 3, p. 3-20.

Kim, D.C., 1980, Recent clay minerals of the Yeongsan Estuary and the adjacent continental shelf: Unpubl. Master's thesis, Seoul National Univ., Seoul, Korea, 63 p.

Kim, H.S., 1971, Metamorphic facies and regional metamorphism of the Ogcheon Belt: Jour. Geol. Soc. Korea, v. 7, p. 221-256.

Kim, J.J., 1970, Recent foraminifera in the Korean Yellow Sea: Rept. Mar. Geol. and Geophys., v. 1, p. 101-118.

Kim, K., 1980, Ocean currents in the southwestern sea off Korea: Unpubl. Tech. Rept., Seoul National Univ., Seoul, Korea, 23 p.

Kim, K.W. and H.K. Lee, 1965, Choongju Geological Atlas: Geol. Surv. Korea.

Kim, N.J., S.W. Kim, and M.H. Lee, 1970, Study on the bottom sediments in the sea area off the west coast of Korea: Rept. Mar. Geol. and Geophys., Geol. Surv. Korea, v. 1, p. 79-99.

Kim, O.J., 1970, Reply to the article "On the geologic age of the Ogcheon Group": Jour. Korea Inst. Min. Geol., v. 3, no. 3, p. 187-192.

Kim, O.J. and J.S. Yoon, 1980, Study on lithologic and tectonic interpretation of the upper Ogcheon Members: Jour. Korea Inst. Min. Geol., v. 13, p. 91-103.

Kim, S.C., 1982, Suspended particulate matters in the Keum Estuary and the adjacent continental shelf: Unpubl. Master's thesis, Seoul National Univ., Seoul, Korea, 69 p.

Kim, S.W., 1973, A study on the terraces along the southeastern coast (Bangeojin-Pohang) of the Korean Peninsula: Jour. Geol. Soc. Korea, v. 9, p. 89-121.

Kim, S.W. and J.H. Chang, 1979, Sedimentological properties of the bottom sediment between Mogpo and Cheongsan Island off the southern coast of Korea: Rept. on Geosci. Min. Resour. KIGAM, v. 5, p. 5-44.

Kim, S.W., J.H. Chang, G.S. Chung, K.J. Cho, C.M. Kim, K.S. Park, W.S. Kim, W.C. Shin, and K.P. Park, 1980a, Marine geological and geophysical survey between Heugsan-Do and Jaeun-Do, southwestern Korea: KIGAM Bull., v. 11, 24 p.

Kim, S.W., C.S. Kim, Y.O. Lee, and S.K. Kim, 1972, Study on the bottom sediments in the middle part of Yellow Sea (part II): Geol. Surv. Korea Bull., v. 14, p. 595-619.

Kim, S.W., Y.O. Lee, and J.H. Chang, 1977, Marine geological investigation of the Asan Bay, west coast of Korea: Rept. on Geosci. and Min. Resour.,

KIGAM, v. 5, p. 163-244.

Kim, S.W. and G.H. Min, 1981, Geological study of the Korean continental shelf between Chin-Do and Cheju-Do, southern coast, Korea: Rept. on Geosci. and Min. Resour., KIGAM, v. 11, p. 75-91.

Kim, S.W., G.H. Min, K.J. Cho, C.M. Kim, K.S. Park, W.S. Kim, and W.C. Shin, 1980b, Geophysical and geological study for base map of marine geology of Korean continental shelf between Jeju Island and Cheongsan Island, southern coast, Korea: Rept. on Geosci. and Min. Resour., KIGAM, v. 9, p. 15-33.

Kim, S.W., Y.S. Park, S.C. Park, K.J. Cho, C.M. Kim, K.S. Park, and W.S. Kim, 1979, Marine geological study of Kyeonggi (Kyunggi) Bay, west coast of Korea: Bull. of Korea Res. Inst. Geosci. and Min. Resour. (KIGAM), p. 1-41.

Kim, W.H. and Y.A. Park, 1981, Microbiogenic sediments in the Nagdong Estuary, Korea: Jour. Ocean. Soc. Korea, v. 15, no. 1, p. 34-8.

Kim, W.Y. and Y.A. Park, 1978, Distribution of trace metals and sediments in estuaries of the Kum River and the Mankyung River: Jour. Ocean. Soc. Korea, v. 13, p. 19-28.

Klein, G., Y.A. Park, J.H. Chang, and C.S. Kim, 1982, Sedimentology of a subtidal, tide-dominated sand body in the Yellow Sea, southwest Korea: Mar. Geol., v. 50, p. 221-240.

Kobayashi, K. and N. Isezaki, 1976, Magnetic anomalies in Japan Sea and Shikoka Basin and their possible tectonic implications, in G.H. Sutton, M.H. Manghnani, and R. Moberly, eds., The geophysics of the Pacific Ocean Basin and its margin: AGU Monogr. Ser, no. 19, p. 235-251.

Kobayashi, K. and M. Nomura, 1972, Iron sulfides in the sediment cores from the Sea of Japan and their geophysical implications: Earth Planet. Sci. Lett., v. 16, p. 200-208.

Koizumi, I., 1970, Diatom thanatocoenoses from the sediment cores in the Japan Sea: Jour. Mar. Geol. Japan, v. 6, p. 1-11.

Koizumi, I., 1978, Neogene diatoms from the Sea of Japan: Mar. Geol., v. 26, p. 231-248.

Koo, J.H., 1972, Marine geophsycial surveys in the northern part of the Yellow Sea: U.N. ECAFE, CCOP Tech. Bull., v. 6, p. 1-12.

Koo, J.H., W. Bosum, and E.G. Kind, 1970, Aeromagnetic survey of offshore Korea: Rept. of Mar. Geol. and Geophys., Geol. Surv. Korea, v. 1, p. 3-40.

Koo, J.H., Y.N. Jang, and J.K. Kang, 1980a, Marine tectonics of Korea: Korea Ocean Res. and Devel. Inst. Bull., v. 2, p. 31-39.

Koo, J.H., J.K. Kang, Y.B. Kim, D.B. Kim, Y.D. Kwon, H.D. Han, and K.K. Kang, 1980b, Marine geology and resources of the Yellow Sea: Korea Ocean Res. Devel. Inst. Bull., BSPE 00023-42-5, 254 p.

Koo, J.H., S.J. Yang, C.M. Kim, W.Y. Lee, W.J. Chun, and Y.D. Kim, 1971, Report on marine geophysical prospecting in offshore area of Seosan district: Rept. of Mar. Geol. and Geophys., Geol. Surv. Korea, no. 2, p. 11-158.

Korea Inst. Energy and Resources (KIER), 1981a, Geol. Map of Korea (1:1,000,000).

Korea Inst. Energy and Resources (KIER), 1981b, Submar. Geol. Map of Contin. Shelf (Series II), II-4, Isopach Map.

Kovylin, V.M. and U.P. Neprochnov, 1965, Structure of earth's crust and sedimentary layer in the central part of the Sea of Japan: Akad. Nuk SSSR Izv. Ser. Geol., no. 4, p. 10-26.

Kozak, L.P., 1974, Distribution of tests of species of planktonic foraminifera in the surface layer of sediments in the Sea of Japan: Geotectonics, v. 11, p. 572-575.

Langseth, M.G., R.V. Huene, N. Nasu, and H. Okada, 1981, Subsidence of the Japan trench forearc region of northern Honshu, in Geology of continental margins, Oceanologica Acta: 26th Inter. Geol. Cong., p. 173-179.

Lee, D.S., 1980, Igneous activity and geotectonic interpretation in the Ogcheon Geosynclinal Zone, Korea—especially referred to ophiolite determination: Yonsei Nonchong, Yonsei Univ., Seoul, Korea, no. 17, p. 109-137.

Lee, G.H., 1983, Distinctive properties of turbiditic and non-turbiditic muds in the Ulleung Basin, East Sea: Unpubl. Master's thesis, Seoul National Univ., Seoul, Korea, 91 p.

Lee, H.Y., 1980, Discovery of Silurian conodont fauna from South Korea: Jour. Geol. Soc. Korea, v. 16, p. 114-123.

Lee, H.Y., M.S. Lee, and S.H. Um, 1980, Geochemistry of amphibolites in the Hwanggangri area, Korea: Jour. Geol. Soc. Korea, v. 16, p. 93-104.

Lee, J.H., 1972, The study of the lower unit of the metamorphic belt in the Ogcheon Geosyncline: Jour. Geol. Soc. Korea, v. 8, p. 25-36.

Lee, J.H., W.J. Chun, Y.D. Kim, and M.I. Koo, 1972, Ground magnetic survey in coastal land area near Kunsan and islands off west coast: Geo. Surv. Korea Bull., v. 14, p. 659-672.

Lee, K., 1979, On crustal structure of the Korean Peninsula: Jour. Geol. Soc. Korea, v. 15, no. 4, p. 253-258.

Lee, K.W., H.S. Kwak, S.H. Lee, and D.S. Lee, 1979, Heavy metals in the Korean coastal waters during summer of 1977: Jour. Oceanol. Soc. Korea, v. 14, p. 1-5.

Lee, M.S., 1981, Geology and metallic mineralization associated with Mesozoic granitic magmatism in South Korea: Mining Geol. v. 31, p. 235-244.

Lee, M.S., in press, Tungsten deposits of Korea, in A.A. Beus, ed., Geology of tungsten: Inter. Geol. Correl. Programme (IGCP), Unesco, Paris.

Lee, M.S. and B.S. Park, 1965, Geological map of Korea, Hwanggangni Sheet: Geol. Surv. Korea.

Lee, S.M., 1974, The tectonic setting of Korea with relation to plate tectonics: U.N. ECAFE, CCOP, Tech. Bull., v. 8, p. 39-53.

Lee, Y.O., 1979, Study of detrital mineral deposits of Jaun, Bigum, Docho and Ui Island: Rept. on Geosci. Min. Resour., Korea Res. Inst. Geosci. and Min. Resour. (KIGAM), v. 5, p. 45-84.

Lelikov, E.P. and I.I. Bersenev, 1975, Early Proterozoic gneiss-migmatite complex of the Japan Sea, southwestern part: Proc. Acad. Sci. U.S.S.R., v. 223, p. 676-679.

Ludwig, W.J., S. Murauchi, and R.E. Houtz, 1975, Sediments and structure of the Japan Sea: Geol. Soc. Am. Bull., v. 86, p. 651-664.

Mammerickx, J., R.L. Fisher, F.I. Emmel, and S.M. Smith, 1976, Bathymetry of the east and southeast Asian seas: Boulder, Co., Geol. Soc. Am. Inc.

Melankholina, Ye.N. and V.M. Kovylin, 1977, Tectonics of the Sea of Japan: Geotectonics, v. 10, p. 273-281.

Meng, C.Y., 1968, Geologic concepts relating to the petroleum prospects of Taiwan Strait: U.N. ECAFE, CCOP, Tech. Bull., v. 1, p. 143-153.

Milliman, J.D. and R.H. Meade, 1983, World-wide delivery of river sediment to the oceans: Jour. of Geol., v. 91, p. 1-21.

Minami, A., 1979, Distribution and characteristics of the sedimentary basin offshore San-in to Tsushima island: Jour. Japanese Assoc. Petrol. Tech. v. 44, no. 5, p. 321-328.

Minato, M., 1973, The origin of deep-basins in the marginal seas behind the island arcs of the western Pacific: Pacific Geol., v. 6, p. 95-100.

Minato, M., M. Gorai, and M. Hunahashi, eds., 1965, The geologic develop-

ment of Japanese Islands: Tokyo, Japan, Tsukiji Shokan, 442 p.
Ministry of Construction, 1974, Report on the Nagdong Estuary: Industrial Site Investigation, Unpubl. Rept., p. 1–56.
Miyake, Y., Y. Sugimura, and E. Matsumoto, 1968, Ionium-thorium chronology of the Japan Sea cores: Rec. Oceanogr. Works Japan, v. 9, no. 2, p. 189–195.
Mizuno, A., H. Sato, and K. Kawamura, 1972, Geological notes on sediment cores from Japan Sea floor: Nihonkai, no. 7, p. 63–72.
Mogi, A., 1979, An atlas of the sea floor around Japan: Tokyo, Japan, Univ. Tokyo Press, 96 p.
Murauchi, S., 1966, Explosion seismology, in 2nd Progress Report on the Upper Mantle Project (UMP) of Japan: Nat'l Comm. for UMP, Sci. Council of Japan, p. 11–13.
Murauchi, S., 1971, The renewal of island arcs and the tectonics of marginal seas, in S. Asano and G.B. Udintsev, eds., Island Arc and Marginal Sea, Tokyo, Japan, Tokai Univ. Press, p. 39–56.
Murauchi, S., T. Asanuma, and K. Hagiwara, 1970, Geological studies on the area off San-in district in the Japan Sea by means of seismic profiler: National Science Museum Rept., v. 13, no. 1, p. 83–90.
Na, K.C., 1972, Regional metamorphism of the so-called Yeoncheon System in the western Gyeonggi Area: Mem. 60th Birthday Prof. C.M. Son, Seoul National Univ., Seoul, Korea, p. 121–140.
Na, K.C. 1980, Regional metamorphism in the Gyeonggi Massif with comparative studies on the Yeoncheon and Ogcheon Metamorphic Belts: Unpubl. PhD thesis, Seoul National Univ., Seoul, Korea, 95 p.
Na, K.C. and D.J. Lee, 1973, Preliminary age study of the Gyeonggi Metamorphic Belt by the Rb-Sr whole rock method: Jour. Geol. Soc. Korea, v. 9, p. 168–174.
Nam, K.S. and Y.H. Seung, 1980, Presentation of current data at Gyema site for the period July, 1979–June, 1980: Korea Ocean Res. and Devel. Inst., project no. 1912-001, 496 p.
Nardin, T.R., B.O. Edwards, and D.S. Gorsline, 1979, Santa Cruz Basin, California borderland: dominance of slope processes in basin sedimentation: Soc. Econ. Paleo. and Miner. Spec. Publ., no., 27, p. 209–221.
National Hydrographic Office of Korea, 1973, Maritime chart no. 240, Approaches to Gamagyang and Yeojaman, Korea.
National Hydrographic Office of Korea, 1980, Tide Table, Yeosu area, Korea.
Niino, H. and K.O. Emery, 1961, Sediments of shallow portions of East China Sea and South China Sea: Geol. Soc. Am. Bull., v. 72, p. 731–762.
Niino, H. and K.O. Emery, 1966, Continental shelf sediments off northeastern Asia: Jour. Sed. Pet., v. 36, p. 152–161.
Niino, H., K.O. Emery, and C.M. Kim, 1969, Organic carbon in sediments of Japan Sea: Jour. Sed. Pet., v. 39, p. 1390–1398.
Normark, W.R. and F.N. Spiess, 1976, Erosion on the Line Islands archipelagic apron: effect of small-scale topographic relief: Geol. Soc. Am. Bull., v. 87, p. 286–296.
Otsuki, K. and M. Ehiro, 1979, Major strik-slip faults and their bearing on spreading in the Japan Sea: in S. Uyeda, et al. eds., Geodynamics of the Western Pacific: Tokyo, Japan, Japan Scientific Soc. Press, p. 537–555.
Park, B.K., 1977, Discussion on the Korean Peninsula, Gondwanaland and Pangaea: Jour. Geol. Soc. Korea, v. 13, no. 2, p. 91–96.
Park, B.K. and I.K. Do, 1973, The Mesozoic granitic batholiths in the Korean Peninsula and new global tectonics: Jour. Geol. Soc. Korea, v. 9, no. 3, p. 149–160.

Park, B.K., S.J. Han, and J.W. Lee, 1976, Clay minerology of bottom sediments in the Jinhae Bay, Korea: Jour. Ocean. Soc. Japan, v. 32, p. 219-227.

Park, B.K., S.J. Han, O.K. Youn, and H.H. Lee, 1976, Recent sediments of Jinhae Bay, Korea: Jour. Geol. Soc. Korea, v. 12, p. 113-123.

Park, B.K. and C.S. So, 1972, The Ogcheon system in the central part of southern Korean Peninsula as an ancient island arc: Jour. Geol. Soc. Korea, v. 8, p. 198-210.

Park, B.K. and M.Y. Song, 1972, A grain size analysis of bottom sediments of Yeonil Bay, Korea: Jour. Ocean. Soc. Korea, v. 7, no. 2, p. 74-85.

Park, K.P., 1982, Trend analysis of aeromagnetic data: Unpubl. Master's thesis, Seoul National Univ., Seoul, Korea, 51 p.

Park, K. P., W.C. Shin, C.S. Kim, K.J. Cho, C.M. Kim, and K.S. Park, 1981, Marine geology and mineral resources of East Sea, Korea (Gangneung-Pohang): Rept. Geosci. Mineral Resour., Korea Inst. Energy and Resour. (KIER), v. 10, p. 149-161.

Park, N.Y., D.S. Kim, H.Y. Choi, and M.H. Lee, 1972, Geology of the western coastal area and islands: Geol. Surv. Korea Bull., v. 14, p. 565-582.

Park, N.Y., S.W. Kim, H.I. Choi, M.H. Lee, and S.K. Kim, 1973, Study on the bottom sediments in the middle part of the Yellow Sea (part III): Report of Geol. and Mineral Explor., Geol. Surv. Korea Bull., v. 1, pt. I, p. 29-50.

Park, Y.A., 1969, Submergence of the Yellow Sea coast of Korea and stratigraphy of the Sinpyeongcheon Marsh, Kimje, Korea: Jour. Geol. Soc. Korea, v. 5, p. 57-66.

Park, Y.A. and M.Y. Song, 1971, Sediments of the continental shelf off the southern coasts of Korea: Jour. Ocean. Soc. Korea, v. 6, no. 1, p. 16-24.

Park, Y.C., 1981, Community structure and spatial distribution of phytoplankton in the southwestern sea of Korea in June, 1980: Unpubl. Master's thesis, Seoul National Univ., Seoul, Korea, 50 p.

Piper, D.J.W., 1978, Turbidite muds and silts on deep sea fans and abyssal plains, in D.J. Stanley and G. Kelling, ed., Sedimentation in submarine canyons, fans and trenches: Stroudsberg, Pa., Dowden, Hutchinson and Ross, p. 163-176.

Prior, D.B. and J.M. Coleman, 1977, Disintegrating retrogressive landslides on very-low-angle subaqueous slopes, Mississippi Delta: Mar. Geotech. v. 3, p. 37-60.

Reeburgh, W.S., 1969, Observations of gases in Chesapeake Bay sediments: Limnol. and Oceanogr., v. 14, p. 368-375.

Reedman, A.J., C.J.N. Fletcher, R.B. Evans, D.R. Workman, K.S. Yoon, H.S. Rhyu, S.W. Jeong, and J.N. Park, 1973, Geological, geophysical and geochemical investigations in the Hwanggangri area, Chungcheong bug-do, Korea: Rept. Geol. Mineral Explor., Geol., Min. Inst. Korea, v. 1, 119 p.

Reedman, A.J. and S.H. Um, 1975, The geology of Korea: Seoul, Korea, Geol. Min. Inst. Korea, 139 p.

Reineck, H.-E. and F. Wunderlich, 1968, Classification and origin of flaser and lenticular bedding: Sedimentology, v. 11, p. 99-104.

Repechka, M.A., 1973, Chemical composition of terrigenous and volcanogenic deep-sea bottom sediments in the Sea of Japan: Geotectonics, v. 10, p. 690-693.

Sakanoue, M., M. Osawa, S. Kitagawa, H. Sugiura, and T. Nakanishi, 1970, Studies on sediment core samples from the Japan Sea by X-ray diffraction, X-ray fluorometry, activation analysis of alpha-ray emitters: Bull. Japan Sea Res. Inst., Kanazawa Univ., p. 75-87.

Schlanger, S.O. and J. Combs, 1975, Hydrocarbon potential of marginal basins bounded by an island arc: Geology, v. 3, p. 397-400.

Schlüter, H.U. and W.C. Chun, 1974, Seismic survey off the east coast of Korea: U.N. ECAFE, Tech. Bull., v. 8, p. 1-15.
Schubel, J.R., 1974, Gas bubbles and the acoustically impenetrable, or turbid, character of some estuarine sediments: Mar. Sci., v. 3, p. 275-298.
Seo, H.J., S.W. Kim, and Y.O. Lee, 1971, Study on the bottom sediments in the middle part of the Yellow Sea (part I): Rept. of Mar. Geol. Geophys., Geol. Surv. Korea, v. 1, p. 69-88.
Shimazu, M., 1979, Green tuff in the core from DSDP Leg 31, Site 302: Kita-Yamatotai: Geol. Mag. v. 85, p. 655-656.
Shitanka, M., F. Ogawa, and W. Ichikawa, 1970, Silicoflagellatae remains in the deep-sea sediments from the Sea of Japan: Nihonkai, v. 4, p. 1-14.
Sibley, D.F. and K.J. Pentony, 1978, Provenance variation in turbidite sediments, Sea of Japan: Jour. Sed. Petrol., v. 48, p. 1241-1248.
Sillitoe, R.H., 1977, Metallogeny of an Andean-type continental margin in South Korea: Implications for opening of the Japan Sea: in M. Talwani and W.C. Pitman III, eds., Island Arcs, Deep Sea Trenches and Back-Arc Basins: AGU Maurice Ewing Monogr. Ser. 1, p. 303-310.
Skornyakova, N.S., 1961, Bottom sediments of the Japan Sea: in Basic characteristics of geology and oceanography of the Japan Sea: Acad. Sci., USSR, p. 23-34 (Nihonkai, no. 2, p. 33-34, translated by Sakanoue, 1968).
So, C.S. and S.M. Kim, 1975, Geochemistry, origin and metamorphism of mafic metamorphic rocks in the Ogcheon geosyncline zone, Korea: Jour. Geol. Soc. Korea, v. 11, p. 115-137.
Son, C.M., 1969, Crustal movement in Korea: Jour. Geol. Soc. Korea, v. 5, p. 167-210.
Son, C.M., 1970, On the geological age of the Ogcheon Group: Jour. Korean Inst. Min. Geol., v. 3, p. 9-16.
Son, C.M., 1971a, Hercynian orogenic cycle in the eastern Asia: Jour. Korean Inst. Min. Geol., v. 4, p. 59-75.
Son, C.M., 1971b, On the Pre-Cambrian stratigraphy of Eastern Asia: Jour. Korean Inst. Min. Geol., v. 4, p. 19-32.
Song, M.Y. and K.J. Cho, 1978, Submarine layer structure by seismic reflection survey between Geoje Island and Namhae Island: Jour. Ocean. Soc. Korea, v. 13, p. 1-8.
Strakhov, N.M., 1962, Principles of Lithogenesis: v. 2, trans. by J.P. Fitzsimmons; ed. by S.I. Tomkeieff and J.D. Hemingway, Consultants Bureau and Oliver and Boyd, New York and Edinburgh, 609 p.
Stroev, P.A., 1971, Gravity anomalies in the Sea of Japan, in Island Arc and Marginal Sea: Tokyo, Japan, Tokai Univ. Press, p. 245-255.
Suk, B.C., 1981, Depositional environment of recent sediment on the Continental shelf around the Jeju Island: Korea Ocean Res. and Devel. Inst. Bull., v. 3, no. 2, p. 123-131.
Tamaki, K., F. Murakami, and E. Honza, 1978, Continuous seismic reflection profiling survey, in E. Honza, ed., Geological investigations in the northern margin of the Okinawa Trough and the western margin of the Japan Sea: Geol. Surv. Japan, Cruise Rept., no. 10, p. 39-42.
Terada, T., 1934, On bathymetric features of the Japan Sea: Bull. Earthq. Res. Inst., v. 3, p. 67-85.
Terman, M.J., 1977, Cenozoic tectonics of East Asia: in M. Talwani and W.C. Pittman III, eds., Island Arcs, Deep-Sea Trenches and Back-Arc Basins: AGU, Maurice Ewing Monogr. Ser. 1, p. 468-470.
Ueno, N., J. Kaneoka, M. Ozima, S. Zashu, T. Sato, and I. Iwabuch, 1971, K-Ar age, Sr isotopic ratio and K/Rb ratio of the volcanic rocks dredged from the

Japan Sea, *in* Island Arc and Marginal Sea, Tokyo, Japan, Tokai Univ. Press, p. 305-309.

Ujiié, H. and M. Ichikura, 1973, Holocene to uppermost Pleistocene Planktonic foraminifers in a piston core from off San-in district, Sea of Japan: Trans. Proc. Paleont. Soc. Japan, no. 91, p. 137-150.

U.S. Naval Oceanographic Office, Pacific Support Group, 1969, Bathymetric Chart of the Sea of Japan: scale 1:2,000,000.

Uyeda, S., 1979, Subduction zones: Facts, ideas and speculations: Oceanus, v. 22, p. 53-62.

Uyeda, S. and H. Kanamori, 1979, Back-arc opening and the mode of subduction: Jour. Geophy. Res., v. 84, p. 1049-1061.

Uyeda, S. and A. Miyashiro, 1974, Plate tectonics and the Japanese island: a synthesis: Geol. Soc. Am. Bull., v. 85, p. 1159-1170.

Wageman, J.M., T.W.C. Hilde, and K.O. Emery, 1970, Structural framework of East China Sea and Yellow Sea: AAPG Bull., v. 54, p. 1611-1643.

Watanabe, T., M.G. Langseth, and R.N. Anderson, 1977, Heat flow in back-arc basins of the western Pacific, *in* M. Talwani and W.C. Pitman III, eds., Island Arcs, Deep Sea Trenches and Back-Arc Basins: AGU, Maurice Ewing Monogr. Ser. 1, p. 137-161.

Workman, D.R., 1972, The tectonic setting of the Mesozoic granites of Korea: Jour. Geol. Soc. Korea, v. 8, p. 67-76.

Yang, S.J., W.Y. Lee, and W.J. Chun, 1971, Report on magnetic survey in Seosan coastal area: Rept. of Mar. Geol. Geophys., Geol. Surv. Korea, no. 2, p. 159-176.

Yasui, M., T. Kishii, T. Watanabe, and S. Uyeda, 1967, Studies of the thermal state of the earth, the 18th paper; Terrestrial heat flow in the Japan Sea (2): Bull. Earthq. Res. Inst., Tokyo Univ., v. 44, p. 1501-1518.

Yasui, M. and S. Uyeda, 1972, Heat flow around Japan: Nihonkai, no. 7, p. 27-38.

Yoshii, T., 1973, Upper-mantle structure beneath the north Pacific and the marginal seas: Jour. Phys. Earth, v. 21, p. 313-328.

Youn, O.K., B.K. Park, and S.J. Han, 1977, Geomorphological evidence of postglacial sea-level changes: Jour. Geol. Soc. Korea, v. 13, no. 1, p. 15-22.

Yuasa, M., K. Tamaki, K. Nishimura, and E. Honza, 1978, Welded tuff dredged from Musashi Bank, northern Japan Sea and its K-Ar age: Jour. Geol. Soc. Japan, v. 84, p. 375-377.

Zenkevitch, N.L., 1959, Submarine topographical map of the Japan Sea (scale 1:2,000,000): Inst. Okeanol., Acad. Sci. USSR.

Zenkevitch, N.L., 1961, Relief of the Japan Sea, *in* P.N. Stepanov, ed., Geological and Hydrological Features of the Japan Sea: Inst. Okeanol. Acad. Sci. USSR., p. 5-22.

Addendum

During the week of my final galley proof, two volumes of reinterpreted seismic and well data on two concession blocks (4 and 6) and two subzones (5 and 7) in the Korean Seas were released (KIER 1980, 1982). The following is a short summary of the additional geological information.

Block 4

The reinterpretation of selected seismic lines made it possible to identify three structural traps as well as a stratigraphic trap. Prominent amplitude anomaly occurs at about 1.0–1.05 s and 1.7 s depths, respectively, both in a gentle anticlinal structure. The Neogene sediment sequence under this structure is more than 3300 m thick. The Plio-Pleistocene unconformity occurs at 530–980 m depth.

Subzone 5

A well (JDZ-V-1) was drilled on an anticline in the subzone 5, a part of the Taiwan Basin in the northern East China Sea (fig. 1.3). The Plio-Pleistocene unconformity occurs at 947 m below the sea floor, underlain by the thick (947–3009 m) Neogene sequences (mudstone, sandstone, shale, and coal beds) that were deposited in shallow marine and lagoonal environments. Acidic tuff beds (3009–3176 m) of unknown age overlie unconformably the basement of Mesozoic biotite granite. The major source rocks favorable for the generation of hydrocarbon occur at 1900–2500 m depth. Potential sandstone reservoirs with favorable anticlinal structures occur at several locations.

Subzone 7

Up to 8 km thick potential sedimentary sequences for hydrocarbon were located in more than thirteen anticlinal structures. A well (JDZ-VII-1) was drilled on one of the anticlines. The Plio-Pleistocene unconformity occurs at 1135 m, below which pre-Eocene nonmarine sequences prevail totaling more than 3800 m in thickness. The acoustic basement of either Cretaceous plutonic or sedimentary rock occurs at about 4900 m at this location.

Block 6

The sedimentary sequence in the Dolgorae-1 well (fig. 1.3) consists of post-Pliocene undeformed strata below about 740 m from the sea floor, overlying the block-faulted Neogene sequence of shallow marine origin. Below 1850 m the sequence is composed of turbiditic mudstone and sandstone (middle to lower Miocene) and the associated mass-flow deposits representing slope facies. Potential reservoirs occur at both 1920-2550 m (late Miocene) and 2800-4262 m (middle Miocene) depths in lithic arenite and lithic graywacke, respectively.

Three stratigraphic traps were located in the central block west of the Dolgorae-1 well composed of channel filling and sigmoidal and oblique progradational sequences of probable slope facies of the paleo-Ulleung Basin. An amplitude anomaly zone was also found near 1.0 s depth.

Miocene clastic sediments prevail in Domi-1 and Sora-1 wells (fig. 1.3) with thin lignite beds interbedded. The occurrence of thick, poorly sorted conglomerates and pebbly sandstone as well as interbeds of mudstone and siltstone suggests an alluvial fan type environment. Pollen study in the Sora-1 well suggests an alternation of continental and shallow marine environments. Potential reservoirs in this well are in the interbedded upper middle Miocene fluvial sandstone about 1300-1700 m deep. In the Domi-1 well it occurs at 1700-2400 m depth, in rocks similar to those in the Sora-1.

Additional References

Korea Institute of Energy and Resources, 1980, Petroleum resources potential in the continental shelf of Korea, No. 80-6, 210 p. [part of block 4 and Korea-Japan Joint Development Zone (JDZ) subzones 5 and 7].

Korea Institute of Energy and Resources, 1982, Petroleum resources potential in the continental shelf of Korea, 342 p. (block 6 and parts of blocks 2, 4, and 5).

Index

Acoustic basement
  Deugryang Bay, 123, 128
  Eastern shelf, 82-84
  Gamagyang Bay, 113, 119
  subzone 7, 152
Aira Caldera, 74, 102, 109
Aira-Tn ash
  East Sea, 74
  Ulleung Basin, 102
Akahoya ash, East Sea, 74
Anmyeon Island, 33
Ash layers
  East Sea, 74
  Ulleung Basin, 102
Aso Caldera, 74, 102
Aso-4 ash
  East Sea, 74
  Ulleung Basin, 102

Baekdo Island, 119
Block 4, 151
Block 6, 152
Bugu Stream, 103

Carbon
  Deugryang Bay, 123
  East Sea, 71
  Ulleung Basin, 100
  Yellow Sea, 42-44
Cenozoic, 134-136
Chlorite
  Deugryang Bay, 123
  East Sea, 69
  Yellow River, 44, 46, 52
Cobalt, Yellow Sea, 46, 51
Copper, Yellow Sea, 46, 51

Deugryang Bay, 111, 122-128
  acoustic characteristics of, 123-128
  Late Quaternary history of, 128, 129
  physiography of, 122-123
  sediments of, 123
Deugryang Island, 123
Dok Island, 66, 89, 103
Domi Basin, 27

East Asia Plate, 134
East China Sea, 4
  subzone 5 of Taiwan Basin of, 151
  surface sediments distribution in, 36, 46

Eastern shelf of Korean Peninsula, 77-88
  acoustic stratigraphy of, 77-84
  geologic setting of, 77
East Sea, 1, 55-75
  acoustic basement of, 58-61, 63-66
  acoustic stratigraphy of, 63-68
  airborne magnetic survey of, 1
  Eastern shelf foraminifera and, 88
  free-air anomaly in, 61-63
  geochemical composition of surface sediment in, 71
  geologic setting of, 4-6, 55-58
  heat flow of, 63
  Holocene-Pleistocene boundary in, 73-74
  Late Quaternary stratigraphy of, 71-74, 136
  lithology of, 71-73
  magnetic and acoustic anomalies of, 61-63
  origin of, 4, 55-58, 61, 133
  paleoceanography of, 74-75
  sedimentary sequence of, 66-68
  seismic survey of, 1-4
  surface sediments distribution in, 69
  velocity structure of, 58-61
  volcanic ash layers of, 74
Eurasian Plate, 134

Foraminifera
  Eastern shelf, 88
  East Sea, 74-75
  Yellow Sea, 42
Fukien Massif, 19
Fukue Basin, 127

Gamagyang Bay, 111-122
  acoustic stratigraphy of, 113-121
  fine sediment deposition in, 121-122
  Late Quaternary history of, 128, 129, 137
  physiography of, 111-113
  sea floor of, 113
  sedimentary structure of, 33, 121, 137
Geological history, 131-137
  Cenozoic, 134-136
  Late Quaternary, 136-137

153

Mesozoic, 133–134
Paleozoic, 132–133
Precambrian, 131–132
Gyeonggi Bay, 44
Gyeonggi Massif, 7–8, 132
Gyeongsang Basin, 7, 12

Heugsan Platform, 19
  acoustic basement of, 28
  sedimentary sequence near, 24
Hino River, 105
Honshu, 58, 66, 68, 69, 105
  Mesozoic period, 133–134
  Ulleung Basin and, 89, 103
Hosan Stream, 103
Hupo Bank, 77
Hupo Basin, 82
Hwangho River, 36, 46, 52, 136
Hydrocarbon
  South Sea, 4
  subzone 5 of Taiwan Basin, 151
  subzone 7, 152
Hyeongsan River, 77
Illite
  Deugryang Bay, 123
  Eastern shelf, 88
  East Sea, 69
  Yellow River, 44, 46, 52
Iron
  East Sea, 71
  Yellow River, 44, 46, 51

Japan Basin, 55
  acoustic basement of, 58–61
  East Sea origin and, 133
  Holocene-Pleistocene boundary in, 74
  magnetic and acoustic anomalies of, 61–63
  sedimentary sequence of, 68
  turbidites in, 103
Japanese Island Arc, 55, 58, 61
Japanese Islands
  Precambrian period in, 132
  East Sea origin and, 61
  Yamato Basin turbidites and, 103
Japan Trench, 55
Jeju Island, 7, 13, 15, 19
  sedimentary basin near, 27
  sediment layer near, 33–36, 40
Jeju volcanic belt, 19
Jinhae Bay, 111, 128

Kaolinite
  Deugryang Bay, 123
  Eastern shelf, 88
  East Sea, 69
  Yellow Sea, 44, 46, 52
Keum Estuary, 51
Keum River, 40, 46, 51, 52
Kikai Caldera, 74
Kita-Yamato Ridge, 55, 58, 63, 89
Korean Peninsula
  Cenozoic events in, 12–13, 134
  Eastern shelf of (see Eastern shelf of Korean Peninsula)
  East Sea origin and, 61
  geology of, 7–13
  Late Quaternary period in, 136
  Mesozoic Orogeny of, 9–12
  Ogcheon Fold Belt of, 8–9
  South Coast embayments of (see South Coast embayments)
  Ulleung Basin and, 89
Korea Plateau, 55
  acoustic basement of, 63, 66
  free-air anomaly near, 63
  geologic setting of, 77
  sediments of, 69
  Ulleung Basin and, 6, 89, 103
Korea Strait
  acoustic stratigraphy of, 82, 84
  Cenozoic period in, 136
  magnetic anomaly in, 19
  paleoceanography of, 75
  sediments of, 40, 69, 88
Koshiki Island, 63
Kula-Pacific Ridge, 55, 133, 134
Kunsan Basin, 24, 27
Kuroshio
  Eastern shelf sediments and, 88
  Yellow Sea foraminifera and, 42
  Yellow Sea sediments and, 36, 40, 52
Kyushu, Japan, 19, 24

Late Quaternary, 136–137
Lead, Yellow Sea, 464, 51
Line Islands, 119

Magnesia
  East Sea, 71
  Yellow Sea, 44
Magnetic anomalies
  East Sea, 61–63
  Yellow Sea, 19

Montmorillonite
  Deugryang Bay, 123
  East Sea, 69
Muds, Ulleung Basin turbidite, 100
Mukho Basin, 82

Nagdong River, 88, 111
Nickel, Yellow Sea, 46, 51
North Kiangsu Basin, 27

Ogcheon Basin
  Paleozoic period, 133
  Precambrian period, 132
Ogcheon Fold Belt, 7, 8-9, 12
  Heugsan Platform as extension of, 19
  Precambrian history of, 132
Oki Bank, 55, 66
  sediments of, 84
  Ulleung Basin and, 89
Oki Island, 63, 103-105
Okinawa Trough, 15, 134
Oki Spur
  minerals in, 105-107, 109
  Ulleung Basin turbidites and, 103

Pacific Plate, 134
Paleoceanography, East Sea, 74-75
Paleozoic, 132-133
Pohang, 68, 77
Pohang Basin, 7, 12-13, 82
Precambrian, 131-132

Samdo Island, 119, 121
Sang Stream, 103
San-In coast, 82, 89, 92, 105, 107, 109
Sea of Japan. *See* East Sea
Sediments, block 6, 152
Sediments, East China Sea, 4
Sediments, Eastern shelf
  acoustic stratigraphy of, 77-84
  surface, 84-88
Sediments, East Sea
  distribution of, 69
  geochemical composition of, 71
  sequence of, 66-68
  surface, 68-71
Sediments, South Coast embayments, 111
  Deugryang Bay, 123
  Gamagyang Bay, 113, 119-122, 137

Sediments, Ulleung Basin, 68, 73, 84, 134-136
Sediments, Yellow Sea, 24-27, 28-36
  clay mineral distribution in, 46
  fine, dispersal of, 52
  isotherm patterns and, 52
  Late Quaternary period, 136
  surface, composition of, 40-44
  surface, distribution of, 36-40
  trace elements distribution in, 46-51
Shanghai Basin, 27
Shantung-Laoyehling Massif, 19
Shimane Peninsula, 89, 103, 105
Sikhote-Alin Range, 61
Sino-Korean Platform, 131
Sobaegsan Massif, 7
Socotra rock, 15, 19
Socotra Subbasin, 27
Song Stream, 103
Sora Basin, 27
South Coast embayments, 6, 111-130
  Late Quaternary history of, 128-129
  sediments of, 111 (*see also* Deugryang Bay; Gamagyang Bay)
South Sea, 1
  acoustic basement of, 19
  airborne magnetic survey of, 1
  Gamagyang Bay and, 111
  magnetic anomaly in, 19
  sedimentary basins in, 27
  sediment composition in, 44
Subzone 5, 151
Subzone 7, 151-152

Taean Peninsula, 46, 52
Taebaeg mountain range, 103
Taebaegsan Basin, 7, 9
Taiwan Basin, subzone 5, 151
Taiwan-Sinzi Folded Zone, 19-24
Tanakura fault, 61
Trace elements, Yellow Sea, 46-51
Tsushima Current
  Eastern shelf sediments and, 84, 88
  East Sea sediments and, 68
  postglacial East Sea sea level rise and, 75
  Yellow Sea sediments and, 40
Tsushima fault, 61
Tsushima Island, 82
Tsushima Straits, 75

Turbidites, block 6, 152
Turbidites, Ulleung Basin, 92, 97–100
    Cenozoic period, 134
    coarse-grained, 97
    factor analysis of, 107–109
    geologic setting of, 103–105
    muds, 100
    provenance of, 103–109
    volcanic ash layers and, 102

Ulleung ash
    East Sea, 74
    Ulleung Basin, 102
Ulleung Basin, 6, 55, 89–109
    acoustic basement of, 58–61, 66
    acoustic stratigraphy of, 89–92
    Cenozoic period, 134–136
    East Sea origin and, 133
    foraminiferal content of, 75
    heat flow in, 63
    hemipelagic facies of, 100–102
    Holocene-Pleistocene boundary of, 74
    Late Quaternary period, 136
    magnetic and acoustic anomalies of, 61–63
    mass-flow deposits in, 92–93
    Mesozoic period, 133, 134
    mineral abundance in, 105–107
    physiography of, 89
    sediments of, 68, 73, 84, 134–136
    stratigraphy of, 102
    turbidite facies of, 97–100
    turbidite provenance of, 103–109
    zoned facies of, 93
Ulleung Interplain Channel, 89
Ulleung Interplain Gap, 89
    hemipelagic facies in, 100–102
    turbidites in, 97, 109
Ulleung Island, 13, 66, 68, 89
    East Sea Ulleung ash from, 74
    geologic setting of, 103
    Ulleung Basin Yamato ash from, 102
Ulleung Seamount, 89

Variscan Orogeny, 58
Volcanic ash layers
    East Sea, 74
    Ulleung Basin, 102

Western Subbasin, 24–27
West Sea. *See* Yellow Sea
West Yellow Sea Basin, 24, 27

Yamato ash
    East Sea, 74
    Ulleung Basin, 102
Yamato Bank, 69
Yamato Basin, 55
    East Sea origin and, 133
    heat flow in, 63
    magnetic and acoustic anomalies of, 61–63
    Oki Spur and, 103
    turbidites, in, 103
Yamato Ridge, 55, 89, 133
    acoustic basement of, 61, 63, 66
Yamato Rise, 69, 71–73
Yangbug Group, 13
Yangtze River, 36, 46, 136
Yellow Sea, 1, 15–54
    acoustic basement of, 15–24, 28
    acoustic wave energy in, 36
    Cenozoic period, 134
    clay mineral distribution in, 46
    fine sediments dispersal in, 52
    foraminiferal content of, 42–44
    geologic structure of, 1, 15–27
    isotherm patterns in, 52
    Late Quaternary period, 136
    Mesozoic period, 133–134
    mid-reflectors and sedimentary layers in, 28–33
    physiography of, 15
    Precambrian history of, 131, 133
    river discharge of sediments into, 36, 40
    sedimentary sequence of, 24–27, 28–36
    sediment dispersal pattern in, 44–54
    shallow geological structure of, 27–36
    surface sediments composition in, 40–44
    surface sediments distribution in, 36–40
    trace elements distribution in, 46–51
    Western Subbasin of, 24–27
Yeongnam Massif, 7, 8, 19, 132
Yeongil Bay, 77
Yeonil Group, 13
Yeongsan Estuary, 44, 46
Yeongsan River, 111
    sediments from, 40, 46, 51
Yungog Stream, 103

Zinc, Yellow Sea, 46–51

About the Author

Sung Kwun Chough, Ph.D., is an associate professor of marine geology at the Seoul National University of Korea. He has participated in numerous geological research projects on the seas surrounding the Korean Peninsula and gathers together here a summary of all existing research for the first time in any language. He serves occasionally as a scientific advisor for the marine geology and geophysics research teams of both the Korea Institute of Energy and Resources and the Korea Ocean Research and Development Institute. He is a member of the American Association of Petroleum Geologists, the American Geophysical Union, the Society of Economic Paleontologists and Mineralogists, the International Association of Sedimentologists, the Geological Society of America, the Geological Society of Korea, the Oceanological Society of Korea, and the Society of Sigma Xi.